# AI, Innovation, and Beyond

Hiroshi Komiyama • Yuya Kajikawa
Cristian Mejia • Yousif Elsamani
Kensuke Fukushi
Editors

# AI, Innovation, and Beyond

## A Report from STS *forum* 2025

Springer

*Editors*
Hiroshi Komiyama
Science and Technology in Society Forum
(STS *forum*)
Tokyo, Japan

Yuya Kajikawa
Institute for Future Initiatives
The University of Tokyo
Tokyo, Japan

Cristian Mejia
Institute for Future Initiatives
The University of Tokyo
Tokyo, Japan

Yousif Elsamani
Institute for Future Initiatives
The University of Tokyo
Tokyo, Japan

Kensuke Fukushi
Institute for Future Initiatives
The University of Tokyo
Tokyo, Japan

ISBN 978-981-95-7019-5 ISBN 978-981-95-7020-1 (eBook)
https://doi.org/10.1007/978-981-95-7020-1

This work was supported by Science and Technology in Society forum.



This Springer imprint is published by the registered company Springer Nature Singapore Pte Ltd.
The registered company address is: 152 Beach Road, #21-01/04 Gateway East, Singapore 189721, Singapore

# Contents

# About the Editors

**Hiroshi Komiyama** is Chairman of the Science and Technology in Society (STS) *forum* and Professor Emeritus at the University of Tokyo. Formerly the university's president (2005–2009), he is a leading voice in sustainability and innovation. Komiyama advocates for integrating science, technology, and policy to address global challenges such as energy, environment, and aging societies.

**Yuya Kajikawa** is a Professor at the Institute for Future Initiatives (IFI), University of Tokyo. He specializes in innovation studies and sustainability science, focusing on technology foresight and knowledge integration. Kajikawa leads projects on energy transitions and circular economy, guiding policy and corporate strategies toward long-term sustainability.

**Cristian Mejia** is a Project Associate Professor at the Institute for Future Initiatives (IFI), University of Tokyo. His research addresses sustainability transitions and global governance, emphasizing evidence-based strategies to achieve the SDGs. Mejia works with international networks to design systemic solutions for climate change, resource management, and socio-economic equity.

**Yousif Elsamani** is a Project Associate Professor at the Center for Global Commons, Institute for Future Initiatives (IFI), University of Tokyo. His work focuses on strategies for managing global commons—climate, biodiversity, and oceans—through interdisciplinary research and policy engagement. Elsamani collaborates internationally to align development with planetary boundaries and foster collective action for sustainability.

**Kensuke Fukushi** is Director and Professor at the Institute for Future Initiatives (IFI), University of Tokyo. His expertise includes urban sustainability, water systems, and resilience planning. Fukushi leads international projects on sustainable cities, advocating for integrated approaches to water, energy, and food systems under climate change.

# Chapter 1
# Introduction: From STS *forum* Dialogue to Global Vision

**Hiroshi Komiyama**

When we gather at the Science and Technology in Society *forum*, we do so with a simple but demanding conviction: science and technology must always be considered in relation to humanity's future. The forum was established to create precisely this space—a place where leaders from public, private, and academic spheres can step back from immediate pressures and examine, together, the long-term lights and shadows of scientific and technological change. That founding purpose remains unchanged. If anything, it has grown more urgent.

Humanity has traveled a very long road. Over hundreds of thousands of years, our species spread across the globe, built civilizations, and gradually improved the conditions of life. Yet for most of history, progress was painfully slow. It was only in the past two centuries—after the Industrial Revolution—that the speed of transformation accelerated dramatically. Science and technology became the principal engines of that acceleration. They lengthened human lives, expanded material prosperity, and created capacities our ancestors could scarcely have imagined. At the same time, they also brought us to a threshold. We now face climate instability, resource constraints, widening social inequality, deepening division, and continuing conflict. Stability can no longer be assumed. The future of our civilization is no longer something that unfolds automatically. It depends on choices we make deliberately and collectively.

This book is written from within that sense of urgency, but it is not written in despair. It is written from the conviction that there is still hope—and that hope lies not in technology alone, but in the ways we connect technology with society, values, institutions, and human purpose. In my opening speech at the STS *forum*, I argued that new foundations for prosperity are emerging: sustainable biomass, renewable energy, and what I have called the urban mine—the vast stock of recoverable materials already circulating in society. Together, these indicate a fundamental shift from

H. Komiyama (✉)
Science and Technology in Society Forum (STS *forum*), Tokyo, Japan



H. Komiyama et al. (eds.), *AI, Innovation, and Beyond*,
https://doi.org/10.1007/978-981-95-7020-1_1

a civilization of extraction to one of circularity. Yet such a transition is not simply a matter of invention. It requires social systems, public participation, institutional redesign, and new forms of shared commitment. That is why the second "S" in STS matters so deeply. Science and technology cannot be separated from society.

A second source of hope lies in knowledge itself. Humanity has accumulated an extraordinary store of scientific, technological, tacit, and cultural knowledge. The problem is not only whether knowledge exists, but whether it is connected and mobilized. Each year at the STS *forum*, I use the image of a porcupinefish as a metaphor for fragmented knowledge: each spine stands apart, as our disciplines and domains of expertise too often do. Knowledge, when disconnected, cannot function as a whole. But if those spines can be connected—if the right combinations of knowledge can be mobilized for the right purposes—then humanity can move forward in new ways. This is where artificial intelligence enters the story. AI, especially generative AI, has the capacity to link knowledge across domains at a speed and scale previously unimaginable. But this very power makes the question of governance unavoidable. AI can help integrate knowledge, yet it can also distort values, divide societies, and magnify existing inequalities if it is misused. It must therefore remain under human control—a servant to human purpose and dignity, never a master.

This conviction animates the present volume. The chapters that follow are not merely records of discussion. They represent an attempt to move from dialogue to direction, from isolated debate to a broader global vision. The sessions collected in this book reflect the STS *forum*'s enduring engagements: first, to harness science and technology for the public good; second, to reform society through dialogue and consensus; and third, to use AI and other emerging tools to connect knowledge and accelerate solutions. These engagements are not abstract ideals. They are practical tasks. They shape how we think about energy, resources, health, inclusion, education, communication, diplomacy, business, and innovation. They also shape how we imagine the future of civilization itself.

The structure of the book reflects this broad ambition.

**Part I, "AI and Beyond,"** begins where many of today's most urgent questions begin: with artificial intelligence. Yet the discussion is not confined to technical capability. The chapters in this section ask what AI means for the next decade of global development, what kinds of roadmaps are needed to guide its deployment, and how major institutions—especially universities—must adapt if they are to remain places where knowledge is created, challenged, and validated responsibly. The chapter on the future university is particularly important in this respect. Universities are among the places where many of the foundational breakthroughs behind AI originated. They are therefore not only users of AI; they are also stewards of its future. The questions raised there—about academic integrity, strategic leadership, cultural diversity, global equity, and public trust—are emblematic of the broader challenge before us.

**Part II, "Sustainability and Society,"** turns to the equally urgent question of how we transition toward a more sustainable civilization. This is not a narrow environmental discussion. It is a civilizational one. A zero-carbon society, a nature-positive economy, and healthier, more resilient human lives all require that we rethink not only technologies, but also institutions, incentives, and ways of living together. The chapter on brain health and brain augmentation may at first seem to belong elsewhere, but in fact it belongs centrally here: sustainability is not only about carbon, materials, and energy; it is also about the conditions for human flourishing, cognition, and dignity in a rapidly changing technological world.

**Part III, "Science, Business, and Policy for Innovation,"** addresses another essential dimension of the transition ahead: how science moves into society through policy, institutions, enterprise, communication, and diplomacy. Fundamental science remains indispensable. Humanity must continue to ask the largest questions—about the universe, life, consciousness, and the deep structure of nature. But science also has a societal trajectory. It enters public life through governance, markets, narratives, and international relations. Innovation cannot be understood only as commercial applications. It must also be understood as the building of societal capacity. The chapters in this section show that science communication and science diplomacy are not secondary activities. They are central to whether knowledge becomes a source of cooperation or division, of trust or mistrust, of shared progress or fragmented advantage.

The concluding chapter returns us to the question with which this volume begins: what kind of future of humankind is possible if science and technology are guided wisely? I believe this is the right closing question, because it reminds us that no chapter in this volume stands alone. AI, sustainability, brain health, innovation policy, business transformation, communication, and diplomacy are not separate conversations. They are interdependent. Our challenge is to connect them.

This, in fact, is the deeper purpose of the STS *forum* itself. We do not convene only to exchange information. We convene to connect knowledge, institutions, and responsibility. In Japan, I have sought to pursue this spirit through the idea of a **Platinum Society**—a sustainable and prosperous social model in which all people can achieve self-actualization. Over the past years, we have identified key sectors in that effort: forests, renewable energy, health, tourism, and human resource development. Across these sectors, the method has been consistent: envision the future, back-cast from 2050, begin with what is feasible now, and scale through institutional learning. Increasingly, AI has become part of that process—not as a substitute for judgment, but as a tool for integrating dispersed knowledge, identifying bottlenecks, aligning actors, and shortening the path from imagination to implementation. This is precisely why AI matters so much in the present historical moment. It is not only that AI is powerful. It is that it can accelerate the work of connection—if we govern it wisely.

Yet acceleration alone is not a virtue. We must ask: toward what end are we moving faster? The answer proposed throughout this book is clear. We must move toward a world that is more circular rather than extractive, more inclusive rather than divided, more cooperative rather than fragmented, and more capable of placing science and technology in the service of humanity. This means linking knowledge across domains, but also linking nations, institutions, and sectors through dialogue and common purpose. It means taking seriously both the promise and the peril of technologies like AI. It means recognizing that the great problems of our time—climate change, inequality, polarization, resource stress, public mistrust—cannot be solved in isolation. They require combinations of knowledge, combinations of actors, and combinations of courage.

For that reason, this volume should be read not simply as a record of expert discussions, but as an invitation to action. The STS *forum* has always aimed to create a place where leaders can think together. But thinking together is not enough. We must also act together. We cannot afford to wait while problems worsen year by year. We must move faster—together. That message shaped the discussions collected in this book, and it remains the spirit in which I offer this introduction. The future is not given to us. It is something we must build—with science and technology, certainly, but also with wisdom, institutions, values, and a shared commitment to the dignity of humankind.

# Part I
# AI and Beyond

# Chapter 2
# AI in 2030 and Beyond

**Yuya Kajikawa**

## 1 The Trajectory of AI toward 2030

> If the last decade showed us that scale unlocks capability, the next may reveal whether coordination unlocks impact—coordination across disciplines and sectors, between human judgment and machine intelligence, and among innovation, governance, and infrastructure.—Yutaka Matsuo

The rapid evolution of artificial intelligence (AI), particularly through the emergence of large-scale foundation models, has already transformed knowledge and creative work in industry and scientific research. The foundation of machine learning models is built using deep neural networks and trained on massive amounts of multimodal data, including text, images, speech, video, music, and motion data. Volume of data and iterations of learning continue to expand, and thus, the performance of learning continues to be improved. As we approach and move beyond 2030, AI systems are expected to transition from assistive tools into deeply embedded, autonomous, and collaborative infrastructures shaping nearly every sector of society and economy. AI has transitioned from a primarily experimental domain into a foundational layer of socio-economic systems.

> AI scientists will fully automate discovery systems that generate hypotheses, run experiments, perform data analysis, and write up results, all of which will be conducted independently of humans.—**Alison Noble**

> We may soon see AI scientists who help process hypotheses and design experiments. If autonomous systems begin to suggest new findings, it will be important to consider validation, auditing, and accountability.—**Yutaka Matsuo**

AI for science has already accelerated hypothesis generation and data analysis. In biotechnology, AI is used to predict genome-wide variant effects and protein

Y. Kajikawa (✉)
Institute for Future Initiatives, The University of Tokyo, Tokyo, Japan



H. Komiyama et al. (eds.), *AI, Innovation, and Beyond*,
https://doi.org/10.1007/978-981-95-7020-1_2

structure, and to integrate multi-omics data and to analyze biological network inference owing to a vast amount of data acquisition by high-throughput experiments (Guo et al., 2025). Autonomous synthesis and characterization are enabled in materials science by integrating robotics and machine learning (Shimizu et al., 2020). In sustainability research, AI is used to predict and reconstruct data for climate change, biodiversity, water resource conservation, energy system analysis, transportation, and smart cities (Nishant et al., 2020).

> Specialized AI systems will be a very common part of our business, government, and daily lives.—**Takayuki Morita**

In business environments, AI is expected to transition from advisory roles in business intelligence, risk management, and strategic planning to autonomous decision-making agents. These agent systems will manage supply chains, dynamically respond to market fluctuations, enter into a contract after negotiation in global markets, optimize team and task structures, and allocate resources with financial strategies. Physical AI, which integrates sensory and cognitive functions with motion control, will augment humans and have an impact on a variety of business sectors, especially in agriculture, construction, transport, and manufacturing industries. Human augmentation technologies contribute not only to the increase of productivity and firm performance but also to enhancing human abilities and reducing the physical burdens of individuals in daily life, to support social inclusion and to decrease the environmental impacts of economic activities (Elsamani et al., 2026).

In the public sector, AI-generated content will increasingly influence public discourse, while there are risks in misinformation and synthetic media, algorithmic amplification of polarization, and erosion of trust. For public service, AI will allow on-demand access to public services, operation, maintenance, and refurbishing social infrastructure, provided deployments consider transparency, fairness, and recourse. It is becoming a tool for politics and elections. For society at large, AI will intersect with education, culture, and democracy. Topics such as misinformation and polarization merit careful attention alongside innovation.

If the last decade showed us that scale unlocks capability, what is the next step for unlocking impacts in 2030 and beyond? For unlocking, we need to consider coordination between human judgment and machine intelligence.

## 2 Artificial Intelligence and Human Intelligence

> What we are talking about—technological singularity, when technology advances faster than human capability—is no longer just theoretical. It's becoming a reality. —**Takayuki Morita**

The rapid advancement of AI has reignited longstanding philosophical and scientific questions about the nature of intelligence itself. AI systems, particularly those based on large-scale machine learning models, now perform tasks once considered exclusive to human cognition, such as language generation, image recognition, and motion control. Singularity, which means a hypothetical future point at which artificial intelligence surpasses human intelligence and becomes capable of improving itself recursively, is a preferred topic for people who want to talk about the future in a way that looks like science.

> What constitutes artificial intelligence for the layman? AI, Artificial General Intelligence, Artificial Narrow Intelligence, Artificial Super Intelligence, or what I just referred to—Broad, Shallow Intelligence. It doesn't matter. There are already computer programs that are smarter than you and me. The question is: Are they useful?—**Randy Goebel**

AI and Human Intelligence (HI) represent two fundamentally different yet increasingly intertwined forms of cognition. While AI systems have demonstrated remarkable capabilities in pattern recognition, optimization, and large-scale data processing, human intelligence remains uniquely characterized by consciousness, intentionality, and contextual understanding. AI imitating HI might not work considering that previous attempts to develop expert systems in the 80s failed. An expert system is a type of AI designed to emulate the thinking and decision-making ability of human specialists. It uses knowledge-based reasoning and rules to infer causality and solve problems in specific domains like machine controls and responses to questions. Despite early success, expert systems often failed because acquiring and maintaining accurate knowledge was difficult and time-consuming. They were also limited by rigid rule-based structures, making them unable to adapt to new situations or handle uncertainty effectively.

Modern AI systems rely heavily on large datasets for training. Learning occurs through exposure to vast quantities of labeled or unlabeled data, enabling models to identify patterns and make predictions. This process is efficient but often brittle, as performance can degrade when faced with data distributions that differ from the training set. Human learning, on the other hand, is highly adaptive and context-sensitive. Humans can learn from minimal examples, conceptualize, transfer knowledge across domains by bridging theoretical and conceptual layers and practical and concrete layers, and incorporate social and cultural information. This form of learning is grounded both in experience, enabling flexible and intuitive adaptation to novel situations, and in logic with strong generalization reasoning capabilities, allowing individuals to apply abstract concepts across diverse contexts. This ability is closely linked to analogy, logic, and the capacity for symbolic thought.

And one of the most profound features of HI is consciousness and ethics. HI is accompanied by subjective experience, self-awareness, and intentionality. These qualities enable reflection, moral reasoning, and a sense of identity. HI includes the capacity for ethical judgment, informed by cultural norms, empathy, and moral principles. Humans can evaluate actions in terms of right and wrong, considering long-term consequences and social impact. AI systems lack consciousness and operate without subjective experience. Their outputs, while sometimes indistinguishable from human-generated content, do not arise from awareness or understanding. AI systems do not possess intrinsic moral understanding. Ethical behavior in AI is externally imposed through design choices, training data, and regulatory frameworks. This raises challenges in ensuring alignment with human values.

> The science of artificial intelligence is about articulating scientific challenges and addressing them. It's not about whether one can write computer programs to behave like humans. I'm the kind of AI scientist who is interested in studying intelligence, not in studying cognitive science. It's not that the study of cognitive science—about how humans are intelligent—is uninteresting. It is interesting. But I first want you to understand that you have to distinguish between the science of intelligence and the scientific challenges that are identified and addressed by AI scientists, rather than those that have to do with your ability to do things.—**Randy Goebel**

Needless to say, AI systems are fundamentally computational constructs, typically based on artificial neural networks and processing architectures. These systems operate through mathematical optimization, adjusting parameters to minimize error functions over large datasets. Their understanding is statistical, derived from correlations within data rather than grounded in real-world experience.

In contrast, human intelligence arises from the biological structure of the brain, consisting of approximately 86 billion neurons interconnected through complex synaptic networks. This architecture supports not only information processing but also consciousness, emotion, and self-awareness. Unlike AI, the human brain integrates sensory, motor, and affective processes called sensorimotor integration. The mammalian cerebral cortex underlies key cognitive processes, including the integration of sensory and motor information, memory formation, and social behavior. Proper brain function depends on a wide variety of specialized cell types, such as neurons, glial cells, and vascular components (Zeisel et al., 2015).

> We need to learn by understanding the trajectory of scientific development. It is the shift from "Consequence Machine" to "Correlative Consequence Machines". Consequence Machine is a computer program whose job is to compute consequences from context that it sees in the world. Early Consequence Machines, like expert systems, were causative consequence machines. They could be given rules from, in the case of the organic chemist or the bacterial diagnostic doctor, that say: "If this is your symptom, this is likely the cause and the disease." That's the truth. In Correlative Consequence Machines, outputs emerge from alleged correlations exposed by relatively shallow compilation of data into opaque black boxes, which you know a lot about as experts on deep learning. From a human viewpoint, you can consider these correlative consequence machines as a computer program—or a human—who has somehow acquired encyclopedic knowledge that is very broad but very shallow.—**Randy Goebel**

In addition to the internal and mechanistic differences of AI and HI, HI is inherently embodied and shaped by physical interaction with the society and environment. Perception, action, and cognition are deeply interconnected, enabling humans to develop intuitive understanding and common sense. AI systems, by contrast, are typically disembodied or only partially embodied in robotic systems. Even in cases of physical AI, interaction with the world is mediated through sensors and predefined objectives, lacking the rich, continuous feedback loops that characterize human experience.

AI and HI differ in their origins, mechanisms, and capabilities. AI represents a powerful extension of computational techniques, capable of surpassing human performance in specific tasks. HI, however, remains uniquely characterized by its integrative, embodied, and conscious nature. Rather than viewing AI and human intelligence as competing forms, it is more productive to consider their complementarity. AI excels in processing scale, speed, and precision. AI supports and augments HI. The integration of HI and AI can enhance efficient and, if properly used, effective decision-making and innovation across domains.

> "As my radiology and AI colleague wrote in one of his paper: 'Will AI replace radiologists?' This is the wrong question. The right answer, he wrote, is: 'Radiologists who use AI will replace radiologists who do not'. This goes far beyond radiology. It applies to lawyers, teachers, software engineers, and even professors. In many knowledge work fields, the people who know how to use AI effectively will replace those who do not. That's why we must learn to design AI systems to augment people, not replace them."—**James Landay**

Humans are existent and moral agents having intention and responsibility. AI systems, however, function as tools or instruments, lacking agency in the philosophical sense. Responsibility for AI actions ultimately lies with developers, operators, and institutions.

For example, in the case of AI application in healthcare, even when AI outperforms HI in detection and diagnosis, we need to answer the following normative questions.

> How do you decide which pathways and conditions to prioritize? What proportion of pathways would benefit from this modernization through a data-driven approach? And how do you generate the trust needed in the clinical profession to adopt totally new ways of treating patients? It will require developing new clinical pathways and new clinical guidelines, which traditionally take many years. We have technical and social complexities surrounding the introduction of AI into the workplace. While there are parallels with other healthcare innovations, the relative novelty of AI at policy, regulation, and service levels creates unique challenges in AI deployment today. This is leading to a new variant of AI called Human-AI Collaboration—or HAIC—which is most easily understood as meaning humans and AI working as a team to make a joint decision.—**Alison Noble**

For unlocking the impact, how should we design coordination between AI and HI? To consider this question, we will consider the design method and process of HAIC and trustworthy AI below.

## 3 Design of Human-Centered and Trustworthy AI

> Indeed, AI is storming into our lives with both promise and peril. The promise is exciting: ChatGPT to answer questions, Gemini and Midjourney to generate images, or Copilot to make us better programmers. We see AI helping with scientific discovery, education, and healthcare—changing all of society, as you will hear from some of the other speakers. But the peril is equally real: AI curating social feeds that spread misinformation and erode democracy; addictive design fueled by AI in gaming and shopping systems; AI systems with embedded racial bias in bail and sentencing systems.—**James Landay**

Trustworthy AI has emerged as a central concept in the responsible development and deployment of AI systems. It refers to AI that is not only technically effective but also aligned with ethical principles, legal requirements, and societal values. This is because AI has raised concerns regarding reliability, fairness, and accountability.

As AI systems increasingly influence critical aspects of society, ensuring their trustworthiness has become essential.

Developing trustworthy AI requires a lifecycle-based approach from requirement specification, architectural design, development of data and trained model, building and testing of the system, and deployment and monitoring (Li et al., 2023; Morley et al., 2020). First, during data collection, high-quality and unbiased datasets must be ensured. Second, model development should incorporate fairness-aware algorithms and robustness testing. Third, deployment must include monitoring mechanisms, human oversight, and risk assessment. Finally, continuous auditing and updating are necessary to maintain system reliability and compliance.

We need a multi-layered governance system from team, organization, industry, and societal levels (Shneiderman, 2020). In addition to reliable AI systems, a safety culture in organizational design, including leadership commitment, talent development, internal reviews, and protocol, is needed. We also need external review and governance systems like industry standards, guidelines, auditing, insurance, and governmental regulations.

> Artificial intelligence is a powerful tool, but it's critical that we design AI systems at the user, community, and society levels. If companies go beyond designing only for their direct users, they aren't more likely to create AI systems with positive societal impact.—**James Landay**

AI has a wide range of applications exerting far-reaching and significant influence. But their algorithms and behavior are complex and opaque, raising the possibility that human oversight may be lost. Furthermore, the conventional development process has been designed to ensure the reliable execution of specific tasks, which may lead to the oversight of unintended consequences. Consequently, behaviors and sometimes the unintended influence of machines are not optimal if their design is based on their own functions and behaviors, and it is necessary to analyze how the interaction between algorithms and humans alters societal outcomes (Rahwan et al., 2019).

Such an analysis on the interaction between algorithms and humans and societal outcomes should have four perspectives (Rahwan et al., 2019). The first is the acquisition of the behavior of AI and machines. The second is the specific environments and conditions triggering behaviors and causal mechanisms generating outputs and outcomes of behaviors. The third is the development of behaviors by data acquisition, learning, system inheritance, and replication. And the last is adaptation to user needs and the environment. With these perspectives, it will become possible to elucidate the influence of machines on humans and society, including the behavior of individual machines, collective behavior of machines, and hybrid behavior between humans and machines.

> Design of human-centered AI has three levels: At the user level, we design the interface, the workflow, and the immediate user experience, and we do this iteratively and test rigorously with our intended users. This is where human-computer interaction and design methods are essential. At the community level, we must consider these broader impacts. Even community-level analysis is not enough. At the society level, we must ask: What happens when the system becomes ubiquitous and has negative effects at the societal scale, as we have seen with Facebook? We must design to mitigate these in advance.—**James Landay**

Let us take the example of rescue drones during disasters to explain this concretely. First, from the perspective of behavior acquisition, conventional methods require behaviors to be pre-set. In contrast, adopting a machine behavior approach provides a perspective for designing functions that enable drones to acquire necessary behaviors on-site to respond to complex or unknown situations within various disaster scenarios. Similarly, from the perspective of mechanisms, this approach raises the new challenge of how to achieve instantaneous learning of such new behaviors in dynamic on-site conditions. From the evolutionary development perspective, individual drones can pass on behaviors learned on-site to other drones, leading to rapid disaster response. Furthermore, from the perspective of adaptation, we can raise questions such as how to enable drones adapted to specific disaster situations to adapt to other situations, and how to set and adapt rescue priorities that vary by region, situation, and stakeholder.

> AI is changing the skills that a human needs to do their job, and in some cases, jobs are becoming obsolete. Healthcare delivery is no exception.—**Alison Noble**

The questions on the coordination between human judgment and machine intelligence are also asking about the new roles and responsibilities of professions. One of the key functions of a profession will definitely be the design of human-centered and trustworthy AI, made with ethical reflections and responsibilities.

## 4 AI Infrastructure, Partnership, and Societal Impact

> Now, one can imagine the next 1,000-fold improvement in performance—from exascale to zettascale.—**Thomas Zacharia**

Enablers of AI are not only the algorithm but also data, computing, and electricity. A defining characteristic of post-2030 AI systems will be their unprecedented scale of data ingestion, storage, and real-time processing. Current frontier models are trained on datasets on the order of $10^{12}$–$10^{13}$ tokens, corresponding roughly to petabyte-scale corpora. However, extrapolation based on observed scaling laws suggests that by 2030 and beyond, AI systems will routinely operate on peta-, exa-bytes datasets.

The nature of data itself is also changing. AI systems are no longer trained solely on static text corpora but will shift toward continuous learning and online adaptation. It increasingly incorporates multimodal, high-resolution, and temporal data streams, including user logs and interactions, videos, IoT signals and sensor data, and outputs from scientific instruments and simulations. For instance, high-quality video datasets alone can require terabytes per hour of raw data, implying that large-scale embodied AI systems (e.g., robotics or autonomous systems) may accumulate exabytes of experiential data annually. Of course, not all data is stored or processed in raw form. Techniques such as data distillation, compression, and retrieval-augmented architectures will reduce storage requirements while maintaining informational richness. Nevertheless, the effective information footprint of AI systems will continue to expand dramatically. These estimates highlight that future AI systems will operate not merely as models trained on large datasets, but as continuously learning entities embedded within planetary-scale data infrastructures.

The computational infrastructure required to support post-2030 AI systems is expected to reach exascale-to-zettascale computing regimes. This implies a transition from today's hyperscale data centers (typically 10–100 MW) to gigawatt-scale facilities. The future data center will be high-density accelerator clusters (e.g., GPUs/TPUs), have advanced interconnects (optical or photonic networking) and specialized cooling systems (liquid or immersion cooling), and be co-located with energy generation sources (e.g., nuclear, renewables). A recent study estimates that overall $CO_2$ emissions in the US could increase up to 0.4–1.9% due to the expansion of data centers by 2030. This increase represents approximately 3–14% of $CO_2$ emissions from the US power sector by 2030 (Jha et al., 2025). AI could become a major industrial-scale energy consumer, comparable to heavy manufacturing sectors. This underscores that the long-term scalability of AI is not merely a computational challenge, but fundamentally an energy systems challenge requiring coordinated advances in infrastructure, efficiency, and sustainable power generation. It is hard for a small or medium-sized company and a university to maintain data center facilities and electricity costs.

> The locus of AI innovation, or at least the practical impact, is really located outside of universities but in private companies. That's because of the tremendous need for computing power and data. Whether that will persist, we don't know, but it's a rare case in the progress of science where for-profit entities are really driving it. It is not the only case—I refer to Shannon and others at IBM Bell Labs—but these tend to be transitory, and then universities are at the forefront of scientific research.—**Daniel Diermeier**

In addition, AI infrastructure in the future does not mean only data and a data center.

> If AI became the nervous system of that infrastructure? What if every bridge, pipeline, and power plant were paired with a living digital twin, continuously analyzing, predicting, and optimizing in real time? Today, intelligence is concentrated in hyperscale data centers. Tomorrow, it will be everywhere, woven into what we believe are three layers: (1) Core Infrastructure: Centralized intelligence, where you will have zettascale systems training frontier models and driving discoveries through other algorithmic advances. (2) Swarm Intelligence: Distributed networks of drones, robots, and edge nodes collaborating in real time. (3) Autonomous Intelligence: Embedded systems acting locally, efficiently, and intelligently. The 1,000-fold breakthrough makes this ecosystem possible. Intelligence moves from a handful of clouds to every bridge, port, hospital, power plant, and pipeline.—**Thomas Zacharia**

AI systems exhibit strong increasing returns to scale. Developing and deploying advanced models requires substantial investments in computational infrastructure, data acquisition, and specialized talent. As a result, mega-firms and well-capitalized institutions are better positioned to exploit AI capabilities and financial returns, which allow further investments. This dynamic leads to capital deepening, where productivity gains accrue disproportionately to owners of capital rather than labor. Firms that successfully integrate AI can achieve significant efficiency advantages, enabling them to outcompete smaller actors and consolidate market share. Over time, this process contributes to market concentration and the emergence of dominant platforms with outsized economic influence.

> AI tends to be a multiplier or an amplifier for those people who are already highly motivated, willing to take chances, and willing to embrace new technologies. We're going to see just a further gap in opportunities emerging as a consequence of that.—**Daniel Diermeier**

AI and its infrastructure have the power to change the current regime and the potential to accelerate social divides and to destabilize societies. AI-driven automation affects labor markets unevenly. Routine and middle-skill jobs are particularly susceptible to automation, while high-skill roles involving creativity, strategy, and oversight are complemented by AI. This results in a polarization of employment, characterized by growth in high- and low-wage jobs and a decline in middle-income occupations. Moreover, AI enables the scaling of high-skill labor. A small number of highly productive individuals or teams, augmented by AI tools, can generate outputs that previously required large workforces. This "superstar effect" amplifies income disparities, as top performers capture a disproportionate share of economic rewards. Such an AI divide accelerates social divides and destabilizes society, and creates the conditions for the emergence of a government that puts the nation first.

In addition, AI development is concentrated in a small number of countries and organizations with access to advanced AI ecosystems, capital, and infrastructure. AI systems can reinforce existing inequalities through feedback loops. For example, biased training data may produce biased outcomes, which in turn generate new data that perpetuates the bias. Similarly, economic advantages gained through early AI adoption can be reinvested to further enhance capabilities, creating path-dependent trajectories of inequality. This creates geopolitical imbalances, as leading nations gain strategic advantages in economic productivity, military capability, and technological sovereignty. It provokes discourse on AI Sovereignty.

> There's another obstacle: power. And I don't mean the electricity that's powering the GPUs. The resources that matter in AI today—compute, data, talent—are concentrated in a handful of companies, and almost all of them are in North America or even Northern California. If we want AI to benefit all of humanity, we can't leave its trajectory to a few corporate actors. We need to band together to help shape AI for societal benefit. These problems are global, so we must collaborate globally.—**James Landay**

AI has the potential to drive significant economic growth and societal benefit, but it also introduces powerful mechanisms for changing the current regime of power and regime. Through capital intensification, labor market polarization, data monopolization, and institutional asymmetries, AI can amplify existing inequalities and create new ones. Addressing these challenges requires proactive policy measures, including investment in education and reskilling, regulation of data and market concentration, and the development of global and inclusive AI governance frameworks. Ultimately, the distributional impact of AI will depend not only on technological progress but on the societal choices that shape its deployment. Then, what is a solution?

> Our idea is simple: pool compute, open-source models, open data, and open software. Build AI resources at scale for the public good with professional software engineering talent. Train the next generation of leaders and researchers and design AI for diverse global contexts. This effort is about redirecting some of the power in the AI market towards human progress. It's about giving universities, governments, and nonprofits the resources to stay in the game. And it's ultimately about steering AI towards science, society, and human well-being.—**James Landay**

> Achieving this vision will require unprecedented public-private partnerships—**Thomas Zacharia**

> Universities are not only educational and research institutions that advance technology, but they're also places where society can reflect on the very policy challenges that were touched upon before.—**Daniel Diermeier**

Within societies, institutions that control AI systems—such as governments and large corporations—may expand their influence over decision-making processes. Algorithmic governance can improve efficiency but also risks reducing transparency and accountability, particularly if systems are opaque or poorly regulated. These asymmetries can lead to a concentration of decision-making power, where critical societal functions are mediated by a limited set of actors. We are at the turning point to unlock AI impact on society by trustworthy and human-centered AI through coordination among innovation, governance, and infrastructure with partnerships across disciplines and sectors.

**Competing Interests** No competing interests.

## Session Panelists

| |
|---|
| **Matsuo, Yutaka (Chair)**<br>Professor, Graduate School of Engineering, The University of Tokyo, Japan<br>Yutaka Matsuo is a professor at Graduate School of Engineering, the University of Tokyo. He received his BS, MS, and Ph.D. degrees from the University of Tokyo in 1997, 1999, and 2002. After working at National Institute of Advanced Industrial Science and Technology (AIST) and Stanford University, he joined the faculty of University of Tokyo in 2007. At Japan Society for Artificial Intelligence (JSAI), he has been a board member from 2020 to 2022. He is the president of Japan Deep Learning Association (JDLA), and a member of the board of directors at SoftBank Group Corp. He is also a member of the Council of New Form of Capitalism Realization under the Prime Minister. He is working on artificial intelligence, especially on deep learning and web mining. |
| **Goebel, Randy**<br>Professor, Faculty of Science - Computing Science, University of Alberta, Canada<br>Citizenship Canadian citizen, born in Melville, Saskatchewan, Canada. Scientific Interests Theory and applications of artificial intelligence, especially knowledge representation, non-deductive reasoning methods, and machine learning, with application to natural language processing, visualization, systems biology, and automated explanation / explainable artificial intelligence, with applications to legal and medical reasoning. Degrees Bachelor of Science (Computer Science) 1974 U. of Regina. Master of Science (Computing Science) 1977 U. of Alberta. Doctor of Philosophy (Computer Science) 1985 U. of British Columbia. Professor of Computing Science, Fellow of the Alberta Machine Intelligence Institute, Adjunct Professor, Faculty of Medicine |

**Landay, James**
Co-Director Stanford Institute for Human-Centered AI (HAI) Anand Rajaraman and Venky Harinarayan Professor in the School of Engineering Professor Computer Science Department Stanford University, Stanford HAI, Stanford University, U.S.A.
James Landay is a Professor of Computer Science and the Anand Rajaraman and Venky Harinarayan Professor in the School of Engineering at Stanford University. He co-founded and is co-director of the Stanford Institute for Human-Centered Artificial Intelligence (HAI). Landay previously was a tenured faculty member at Cornell Tech, the University of Washington, and UC Berkeley. He was also Director of Intel Labs Seattle and co-founder of NetRaker. While on sabbatical at Microsoft Research Asia in Beijing, he taught for one semester at Tsinghua University. Landay received his BS in EECS from UC Berkeley, and MS and PhD in Computer Science from Carnegie Mellon University. He is a member of the ACM SIGCHI Academy, an ACM Fellow, and was awarded the ACM SIGCHI Lifetime Research Award.

**Noble, Alison**
Foreign Secretary and Vice-President, The Royal Society; Technikos Professor of Biomedical Engineering; Fellow of St Hilda's College, Department of Engineering Science, University of Oxford, U.K.
Professor Alison Noble CBE FREng FRS is currently the Oxford University Technikos Professor of Biomedical Engineering, and a Vice President and Foreign Secretary of the Royal Society. Alison's research interests are at the inter-disciplinary interface of artificial intelligence (computer vision) and healthcare imaging. She has worked with clinical partners in the UK and worldwide on translational AI-based imaging science and commercialised some of her group's research, leading to technology now used worldwide in GE ultrasound machines. In addition to her Foreign Secretary role, Alison has chaired Royal Society science policy working groups leading to reports on "Protecting privacy in practice" (2019), "From privacy to partnership" (2023) and most recently "Science in the age of AI" (2024).

**Morita, Takayuki**
President and CEO (Representative Executive Officer), Member of the Board (Member of the Compensation Committee), NEC Corporation, Japan
Takayuki Morita became President and Chief Executive Officer (Representative Director) of NEC Corporation, a leader in the integration of IT and network technologies, on April 1, 2021. Previously, he was Chief Financial Officer, where he managed the financial strategy and operations of NEC's 100,000 employees and total revenue of US$25 billion. He was also responsible for Corporate Business Development functions, including the oversight of M&A and other investments, such as major advancements in the digital finance, digital government and drug development fields. In the past twenty years, he has played a central role in the development of partnerships and alliances with corporations across the globe having completed more than forty deals of corporate mergers, acquisitions and investments.

**Diermeier, Daniel**
Chancellor, Vanderbilt University, U.S.A.
Daniel Diermeier is the chancellor of Vanderbilt University. He is a fellow of the American Academy of Arts and Sciences and the Guggenheim Foundation. He has published five books with a sixth forthcoming and more than 100 research articles. Before coming to Vanderbilt, Diermeier served as the provost of the University of Chicago, dean of the Harris School of Public Policy. Prior to that, he was a professor at the Kellogg School of Management at Northwestern University. Diermeier earned a Ph.D. in political science from the University of Rochester. He also holds master's degrees in political science from the University of Rochester and the University of Munich, and a master's degree in philosophy from the University of Southern California.

**Zacharia, Thomas**
Senior Vice President, Strategic Technology Partnerships and Public Policy, AMD, U.S.A.
Dr. Thomas Zacharia is senior vice president of strategic technology partnerships and public policy for AMD. He leads the global expansion of AMD public/private relationships with governments, NGOs and other organizations to help fast-track the deployment of AMD-powered AI solutions. He is a distinguished industry leader with extensive experience in scientific research, technology development, public policy and strategic public/private partnerships. He has provided expert testimony to US Congress in his previous role as director at Oak Ridge National Lab. Dr. Zacharia holds a BS in mechanical engineering from the Natl Inst. of Technology Karnataka, Surathkal, India, an MS in materials science from the Univ. of Mississippi, and a PhD in engineering science from Clarkson Univ. in Potsdam, NY.

## References


Elsamani, Y., Mejia, C., & Kajikawa, Y. (2026). Mapping human augmentation technologies for societal impact: A multilevel framework for classification and innovation potential. PLoS One, 21(2), e0343292.

Guo, F., Guan, R., Li, Y., Liu, Q., Wang, X., Yang, C., & Wang, J. (2025). Foundation models in bioinformatics. National Science Review, 12, nwaf028.

Jha, R., Jha, R., & Islam, M. (2025). Forecasting US data center CO2 emissions using AI models: Emissions reduction strategies and policy recommendations. Frontiers in Sustainability, 5, 1507030.

Li, B., Qi, P., Liu, B., Di, S., Liu, J., Pei, J., Yi, J., & Zhou, B. (2023). Trustworthy AI: From principles to practices. ACM Computing Surveys, 55(9), 177.

Morley, J., Floridi, L., Kinsey, L., & Elhalal, A. (2020). From what to how: An initial review of publicly available AI ethics tools, methods and research to translate principles into practices. Science and Engineering Ethics, 26, 2141–2168.

Nishant, R., Kennedy, M., & Corbetta, J. (2020). Artificial intelligence for sustainability: Challenges, opportunities, and a research agenda. International Journal of Information Management, 53, 102104.

Rahwan, I., Cebrian, M., Obradovich, N., Bongard, J., Bonnefon, J. F., Breazeal, C., Crandall, J. W., Christakis, N. A., Couzin, I. D., Jackson, M. O., Jennings, N. R., Kamar, E., Kloumann, I. M., Larochelle, H., Lazer, D., McElreath, R., Mislove, A., Parkes, D. C., Pentland, A. S., Roberts, M. E., Shariff, A., Tenenbaum, J. B., & Wellman, M. (2019). Machine behaviour. Nature, 568(7753), 477–486.

Shimizu, R., Kobayashi, S., Watanabe, Y., Ando, Y., & Hitosugi, T. (2020). Autonomous materials synthesis by machine learning and robotics. APL Materials, 8(11), 111110.

Shneiderman, B. (2020). Bridging the gap between ethics and practice: Guidelines for reliable, safe, and trustworthy human-centered AI systems. ACM Transactions on Intelligent Systems, 10(4), 26.

Zeisel, A., Muñoz-Manchado, A. B., Codeluppi, S., Lönnerberg, P., La Manno, G., Juréus, A., Marques, S., Munguba, H., He, L. Q., Betsholtz, C., Rolny, C., Castelo-Branco, G., Hjerling-Leffler, J., & Linnarsson, S. (2015). Cell types in the mouse cortex and hippocampus revealed by single-cell RNA-seq. Science, 347(6226), 1138–1142.

# Chapter 3
# Global Impact of AI by 2030 and Beyond: Roadmap for the Next Decade

**Yuya Kajikawa**

## 1 The Light and Shadow of AI in the Economy and Society

> AI is probably more light than shadow. Light connotes hope, and definitely we see that in AI.—**Li-Na Koh**

Artificial Intelligence (AI) represents one of the most transformative technological advancements of the modern era, offering a wide range of advantages across economic, scientific, and societal domains. Its core strength lies in its ability to process vast amounts of data, identify patterns, and execute tasks at speeds and with precision far beyond human capability. As such, AI is not merely a tool for automation, but a general-purpose technology that enhances productivity, augments human intelligence, and enables entirely new forms of innovation.

> One of the most attractive features of AI is that the models run faster and with less computational resources. So, for a 10-day weather forecast, for example, if I were to run a numerical weather prediction model, it would typically take a few hours, but with an AI emulator or surrogate, I can do it in minutes. When we did our national climate change projections for Singapore, which we launched last year, we took four years—four years—to downscale the global climate models to a regional projection set and further downscale that for Singapore. With AI, I can do that much faster.—**Li-Na Koh**

One of the primary advantages of AI is its scalability. Once developed, AI systems can be deployed across millions of users and applications at relatively low marginal cost. This allows organizations to deliver consistent and high-quality services at scale. Another key benefit is enhanced efficiency and productivity. AI can automate repetitive and routine tasks at lower cost, freeing human workers to focus on higher-value activities such as creative problem-solving and strategic decision-making. AI is even effective in tasks. AI systems can replace human experts by producing more creative works and by reducing the likelihood of accidents caused by human error.

Y. Kajikawa (✉)
Institute for Future Initiatives, The University of Tokyo, Tokyo, Japan



H. Komiyama et al. (eds.), *AI, Innovation, and Beyond*,
https://doi.org/10.1007/978-981-95-7020-1_3

The superiority of AI and AI-supported systems is becoming apparent when task structure and human-AI interactions are well designed (Vaccaro et al., 2024). AI also excels in data-driven decision-making, or at least informed decision-making and decision support. By analyzing large and complex datasets, AI systems can uncover insights that would be difficult or impossible for humans to detect. AI has considerable potential to enhance transparency and reduce bribery by limiting human control and corruption in administrative processes such as public procurement and budgeting systems (Allahrakha, 2025).

But these advantages also cause other problems. The scalability of AI systems yields high returns on investment and is often monopolized by leading companies. Improved efficiency and effectiveness can widen economic divides due to differences in accessibility and availability across individuals, communities, organizations, countries, and regions. In data-driven decision-making, there is a risk of biased input data, hallucinated outputs, and malicious tampering.

> AI is not universally accessible, only for well-resourced countries. But I have a slightly different take from the AI on this. AI has made the science very accessible.—**Li-Na Koh**

> Digital divide and AI divide, having the three key dimensions of access, usage, and quality, will emerge especially in the Global South because of the lack of reliable internet connectivity, affordable devices, and stable electricity. In essence, the world is moving at two speeds: one digitally accelerated Global North, and a digitally constrained Global South. This digital divide has consequences that can lead to an unequal AI future. So if we leave these issues unaddressed, the divide could lead to AI colonialism, where data infrastructure and benefits are controlled by just a few powerful nations or corporations.—**Bolanle Ojokoh**

> Some of the likely consequences of AI are: (1) Widening economic inequality: AI-driven automation could increase productivity and GDP in advanced economies, while in low-income economies, AI is displacing jobs for some people. (2) Educational disparities: Students in resource-poor regions face the risk of being excluded from AI-enhanced learning, perpetuating generational inequality. (3) Data dependency: Countries that don't have enough data infrastructure will always have to rely on imported AI models that are not trained on their own data, making it challenging for such models or systems to respond to the particular context in which they are being used. (4) Ethical imbalance: Ethical AI principles and governance frameworks are mostly shaped in the Global North, and this will also be a challenge if they are to be applied in the Global South.—**Bolanle Ojokoh**

However, compared with the empirically obvious light side of AI, the shadow of AI remains vague. We usually hear people talking about hallucination and bias, but we humans are also sometimes wrong and biased. The shadow side of AI has not been systematically analyzed, leading to the broad perception that AI is probably more light than shadow. But even if we have a net gain by implementation and dissemination of AI, resolving the negative sides will lead to maximizing the positive impact of AI on society. Nevertheless, AI's shadow side cannot be neglected, as discussed below.

## 2 Cybersecurity and National Security Threats Posed by AI

> In the same way that drones in recent warfare have changed the landscape of physical warfare, dark AI presents a set of threats the likes of which we've yet to see. In fact, on the dark AI web today, you can buy for less than $5 a tool that will allow you to write a first-time, never-seen attack that can make nation-states vulnerable.—**Lane Bess**

> AI leadership is a critical component of international competition and cooperation. Countries are increasingly worried that hostile state actors, through espionage, malicious use, or export of risky AI, could undermine national security or global stability.—**Maria T. Zuber**

Some aspects of AI's light and shadow should be investigated to determine whether they are net positives or negatives. But we have other issues we should avoid or hesitate to evaluate by net effect, which are difficult to quantify and sometimes irreversible and serious. Examples of the latter include AI-enabled cybersecurity and national security threats.

AI could pose new cybersecurity challenges. Traditional cybersecurity threats include malware, backdoors, phishing, and insider threats. AI will amplify cybersecurity risks by enabling polymorphic malware that adapts its behavior to evade detection, generating highly personalized and convincing messages (e.g., deep contextual phishing), optimizing DDoS attacks, and discovering vulnerabilities via automated code analysis. Additional threats originating from AI include deepfake impersonation, dissemination of disinformation, bypassing biometric authentication, data poisoning, model inversion and theft, privacy leakage, and autonomous cyberattacks and hacking (Gupta et al., 2023).

In addition to cybersecurity and cybercrime, AI can cause cross-national conflicts that threaten national security, as quoted below.

> (1) Scaled influence and deception: For example, deepfake news, especially around elections. (2) Faster and broader cyber operations: AI boosts reconnaissance, phishing, vulnerability discovery, and malware development. (3) Model theft and dual-use R&D: Adversaries will target labs, cloud providers, and integrators to steal model weights, training data, or code. They'll also probe frontier models for abusive capabilities such as dual-use bio and cyber applications. (4) Sanctions evasion and illicit finance: AI-augmented impersonation and job fraud, including automated applications and interviews, identity fabrication—these are helping sanctioned regimes earn revenue and gain access inside firms, creating both security and compliance exposures. (5) Military and security uses below open conflict: Expect broader use of AI for ISR, triage, target recognition, cyber support, and autonomous or semi-autonomous systems, plus the possibility of domestic surveillance and repression.—**Maria T. Zuber**

## 3 Regulatory Framework of AI

How should we handle the shadow and dark sides of AI? One is regulation; the other is sovereignty. Regulation and governance are related but distinct concepts. Regulation refers to formal rules and laws established by governments or authorities to control behavior and ensure compliance through enforcement mechanisms. In contrast, governance is a broader concept that encompasses the processes, structures, and norms through which organizations or societies are directed and managed. Governance includes not only regulations but also voluntary guidelines, ethical principles, and internal policies. While regulation is typically mandatory and externally imposed, governance can be both formal and informal, involving multiple stakeholders. Thus, regulation is a component of governance, but governance extends beyond legal control to include coordination, accountability, and decision-making frameworks. These approaches are illustrated in the following quotation.

> (1) Risk-based regulatory frameworks: Drawing lines between low-risk and high-risk uses. High-risk uses get much stricter oversight, and some uses are actually going to be banned outright. (2) Content labeling and disclosure: Regulating generative AI content by requiring services to label AI-generated or synthetic content so that users know it's generated. (3) Licensing and registration: The requirement that providers of certain public or general-use AI systems obtain licenses and register their models. (4) Prohibiting or restricting certain practices: Bans on things like social scoring and emotion detection in sensitive areas such as workplaces or schools, covert manipulation, and certain surveillance uses. (5) Transparency and accountability mechanisms: Requirements for reporting safety incidents, disclosing risk mitigation processes, and ensuring human oversight. (6) And finally, governance institutions and enforcement: Setting up authorities or offices that will monitor compliance and do market surveillance.—**Maria T. Zuber**

There are opposing policies on AI governance. One is proactive, aiming to export AI and its values worldwide without regulations, while the other is conservative, strengthening regulations in privacy protection and risk assessment. The US and China are typically the former, and the EU tends to be the latter (Cohen et al., 2020). We can also analyze state policy along the four dimensions of the entrepreneurial state (like China), the market-oriented state (like the USA), the self-regulation-promoting state (like Japan), and the regulatory state (like Norway). At the same time, each country has multiple policy directions (Djeffal et al., 2022). The different policy directions among nations and regions make harmonizing the global regulatory framework difficult, and international AI policy initiatives for regulation and governance are fragmented (Cihon et al., 2020).

> AI governance is shaped by both collaboration and competition. Cooperative forums—this was mentioned in the talk before mine—such as OECD, for example. Competitive dimensions: nations seek leadership in AI for economic and security reasons, sometimes leading to divergent rules. Alignment efforts: talks are underway on shared standards for testing frontier models and managing catastrophic risks. These efforts reflect a shift from purely domestic regulation of AI to geopolitical AI governance, treating powerful models and hardware as strategic technologies. The effectiveness depends on closing loopholes, sustaining international cooperation, and managing the tension between open innovation and security-driven restrictions. Now, in the future of governance, I think over the coming five years we might see a structure that spans both regional regulation and market-driven incentives. And finally, there are going to be continued tensions between innovation and control, especially in the US, where they're emphasizing very permissive innovation, whereas in the EU and global institutions, they're pushing for more precautionary norms. And competitors will execute internal controls but will be permissive of uses against adversaries.—**Maria T. Zuber**

If we aim to introduce regulations, despite their difficulty, we should avoid prescriptive regulations and adopt performance-based ones to encourage technological development and innovation (Nakamura & Kajikawa, 2018). Good regulation promotes innovation. If safety is not efficiently regulated, companies refrain from investing due to excessive limitations and requirements. If safety is not effectively regulated, companies also refrain from investment due to market uncertainty. By demonstrating the desired performance status under the regulation, resources are invested in developing technologies and meeting the criteria.

## 4 AI Sovereignty

Sovereignty refers to the supreme authority of a state to govern itself without external interference. It includes control over territory, laws, resources, and political decision-making. A sovereign state has the power to establish and enforce its own legal system, regulate economic activity, and conduct foreign relations independently. Traditionally, sovereignty is associated with the nation-state system, where each state is recognized as equal under international law. In practice, however, sovereignty can be influenced by globalization, economic interdependence, and international agreements, which may limit absolute autonomy while preserving formal independence.

> The global pandemic a few years ago taught every country, every government, a very, very important lesson. At a time when there were shortages of N95 masks, food supply shortages, digital chip shortages, and vaccine shortages, it became clear that the control, reliability, and availability of your supply chain were of critical national importance. In other words, sovereignty. And there are five areas that were absolutely non-negotiable: food sovereignty, health sovereignty, energy sovereignty, digital services sovereignty, and, of course, national defense. It's obvious that AI sovereignty is of critical national importance, and it's a non-negotiable thing.—**Amit Mital**

AI sovereignty refers to a nation's ability to independently develop, deploy, and govern artificial intelligence technologies in alignment with its own laws, values, and strategic interests. It includes control over key components such as data, algorithms, operating systems, devices, and computational infrastructure. AI sovereignty also implies reducing reliance on foreign technologies or platforms that could create vulnerabilities or dependencies. This concept has gained prominence as AI becomes a critical driver of economic competitiveness, national security, and social governance, making technological autonomy increasingly important for states.

AI sovereignty is important because it enables countries to maintain control over critical digital infrastructure and protect national interests. Without it, nations may depend on foreign AI providers, raising risks related to data privacy, security, and political influence. It also supports economic resilience by fostering domestic innovation and reducing technological dependency. Furthermore, AI sovereignty allows governments to enforce regulations and ethical standards consistent with local values. As

AI systems shape decision-making across healthcare, defense, and public services, maintaining sovereignty ensures accountability, trust, and long-term strategic stability.

> What does AI sovereignty actually mean? It means that the system that is going to impact every citizen in my country has to be a system that I have control over, transparency over, visibility over—that no other hostile power can turn it off, can influence it, can change the results I see by poisoning the results to further adversarial impact.—**Amit Mital**

The narrative supporting AI sovereignty sounds plausible at first glance. But we need to be careful. We need to consider fundamental questions. What does AI sovereignty mean? For whom is it needed? And to whose benefit? A vague conceptualization of AI sovereignty could be used to pursue diverse, even conflicting, goals. Recent research (Mügge, 2024) argues that the EU's AI policy places greater emphasis on maintaining its own regulatory autonomy than on empowering individual citizens, reflects a competitive global race mindset in AI development, and pays insufficient attention to its effects on populations outside the EU. This approach to AI sovereignty both mirrors and contributes to increasing geo-economic tensions.

## 5 Global Governance and Collaboration

Introducing regulation is practically difficult, while good regulation may promote innovation. It is also unrealistic for each country to have sovereignty at every layer of data, algorithms, operating systems, devices, and computational infrastructure. A vague conceptualization of AI sovereignty distorts discourse and hampers detailed and relevant design. More important is global governance and global collaboration. And the global governance framework needs to involve diverse but related stakeholders, including academia, service providers, international governmental organizations, governmental institutions, legislative and judicial bodies, and society (de Almeida et al., 2021).

> Looking towards 2030, in terms of trying to chart a roadmap for inclusive and equitable AI, we need to seek to create a globally inclusive AI ecosystem. In the next decade, we must prioritize bridging these gaps. How do we bridge the gap in terms of access to infrastructure, literacy, and governance? And how do we enhance collaboration between the North and the South?—**Bolanle Ojokoh**

> These efforts reflect a shift from purely domestic regulation of AI to geopolitical AI governance, treating powerful models and hardware as strategic technologies. The effectiveness depends on closing loopholes, sustaining international cooperation, and managing the tension between open innovation and security-driven restrictions.—**Maria T. Zuber**

According to the discussion in the session, global collaboration is needed in the following areas, among others:

(1) Foundational AI Infrastructure and Data Commons

> How do you curate a world of data? The answer may be that you don't. The answer may be that you go to some derivative of AI, such as agentic AI, where you pick a very finite dataset that you can curate, and then you get very, very accurate solutions.—**Bernard S. Meyerson**

> We have very big data, and the World Meteorological Organization (WMO) is, in fact, one of the best examples in the UN system of global data sharing.—**Li-Na Koh**

> On global health using AI, the very first question that came up was: 'I can't get the data.' "It can't go out of the country. It can't be personally identifiable."—**Bernard S. Meyerson**

> We need to build foundational infrastructure by investing in broadband expansion and maybe regional AI hubs in developing economies like in Africa, Asia, and Latin America, in order to ensure that they have localized innovation and infrastructure that can serve their specific contexts. There is also the need for public-private partnerships.—**Bolanle Ojokoh**

Foundational AI Infrastructure and Data Commons are necessary to address structural imbalances in the AI ecosystem and to ensure that the benefits of AI are broadly distributed. Currently, access to large-scale computing resources and high-quality data is concentrated in a small number of technology companies and well-funded institutions. This concentration creates barriers to entry, limits competition, and risks reinforcing economic and technological inequalities. By establishing shared infrastructure and data commons, societies can democratize access to critical AI resources. This enables a wider range of actors to participate in innovation, fostering diversity, inclusion, and equity of development, access, and public use of AI in healthcare, climate services, education, and other public domains.

Additionally, data commons facilitate responsible data governance. By implementing standardized frameworks and data-sharing protocols for privacy protection and accountability, they help mitigate risks related to misuse, bias, and fragmentation. This is particularly important in domains where data sensitivity and ethical considerations are paramount. From a strategic perspective, foundational infrastructure enhances resilience and shared sovereignty. Nations and regions can reduce dependence on external providers and ensure secure, reliable access to AI capabilities. At the same time, shared systems can promote international collaboration by enabling interoperable and trusted data exchanges.

(2) Education and Capability Building

> It's important to foster digital and AI literacy, because in some nations, AI is still not being integrated into the curricula right from basic education. There is a need to do that so that everybody progresses together and there is no marginalization, both among the youth, women, and even underserved communities.—**Bolanle Ojokoh**

A global, cross-border education system in AI, driven by Global South–Global North collaboration, is essential to address both inequality and capacity gaps in the global AI ecosystem. Currently, advanced AI research, infrastructure, and expertise are concentrated in a small number of countries in the Global North. In contrast, many countries in the Global South face limited access to training, resources, and opportunities. This imbalance risks deepening technological dependency and excluding large populations from participating in AI-driven innovation.

Collaborative education systems—such as joint degree programs, researcher mobility, and shared curricula—can help distribute knowledge more equitably and build local capacity. They also enable mutual learning: the Global North contributes technical expertise and infrastructure, while the Global South offers diverse data, perspectives, and problem contexts that are critical to developing inclusive, globally relevant AI systems.

Furthermore, such partnerships strengthen global resilience by creating a more geographically distributed talent base. This reduces concentration risks and supports more diverse innovation pathways. In an interconnected world, shared challenges such as climate change, public health, and sustainable development require AI solutions informed by global perspectives. Therefore, cooperative talent development is not only a matter of equity but also a prerequisite for effective and responsible AI at scale.

(3) Collaborative research on AI governance and policy, and harmonization among state policies

> Data sovereignty and inclusive research are also important. Nations should be empowered to develop data governance policies that will protect privacy while encouraging local innovation, and to promote cross-regional data collaborations where Global South researchers can contribute to what happens in the global context, because we have a lot of talents and initiatives that these people can drive.—**Bolanle Ojokoh**

> Ethical AI and global governance is another area. There is a need to establish multilateral AI governance frameworks that ensure fairness, transparency, and accountability across borders. It's important also to include voices from the Global South in global AI standard-setting bodies such as UNESCO, OECD, and United Nations AI advisory initiatives.—**Bolanle Ojokoh**

Collaborative research on AI governance and policy, along with harmonization among state policies, is essential because AI systems operate across borders while regulations remain largely national. Without coordination, fragmented rules can create regulatory gaps, inconsistencies, and arbitrage opportunities, allowing harmful uses of AI to persist while increasing compliance burdens for organizations operating globally. Collaborative research also enables countries to share knowledge, align risk assessments, and develop common standards for safety, transparency, and accountability. It also contributes to the development of AI researchers. This is particularly important for managing high-risk applications and frontier AI systems, where unilateral approaches are insufficient to address global impacts.

Policy harmonization also facilitates innovation by providing a cleaner, more predictable regulatory environment. Companies and researchers benefit from interoperable rules, which reduce uncertainty and enable cross-border collaboration. At the same time, coordinated governance helps ensure that AI development reflects shared values, such as human rights, fairness, and security. Ultimately, collaborative governance strengthens global stability by balancing competition with cooperation. It allows states to collectively address transnational risks such as misuse, cybersecurity threats, and misinformation while supporting responsible innovation in an interconnected AI ecosystem.

**Competing Interests** No competing interests.

## Session Panelists

| |
|---|
| **Meyerson, Bernard S. (Chair)**<br>Dr. Meyerson serves as IBM's Chief Innovation Officer Emeritus. Joining IBM in 1980, Dr. Meyerson invented Silicon-Germanium technology, founding IBM's highly successful Mixed Signal business. As an executive, he has led IBM's global semiconductor development, strategic alliances for the Systems and Technology Group, becoming IBM's Chief Innovation Officer in 2010. His team leads the definition and integration of corporate-wide technical and business strategic initiatives. Dr. Meyerson is a Fellow of the American Physical Society, the IEEE, and a member of the US National Academy of Engineering. Most recently, Dr. Meyerson was honored with the 2015 National Medal for Public Service by the government of Singapore for his many years of service and impact as a technical advisor to the nation. |
| **Koh, Li-Na**<br>Ms Koh Li-Na is the Deputy CEO responsible for meteorological services, weather and climate science research, and radiation protection at the National Environment Agency of Singapore. She is also the Director-General of Meteorological Services. Prior to NEA, she was the Assistant CEO at the Infocomm Media Development Authority of Singapore, overseeing the formulation of strategies for Singapore's digital economy and digital inclusion for society. Ms Koh has also served stints at the Early Childhood Development Agency, the Info-communications Development Authority and the Personal Data Protection Commission. Ms Koh holds a Master in Business Administration (Distinction) from INSEAD and a Master in Engineering (Distinction) from Cambridge University. |
| **Ojokoh, Bolanle**<br>Bolanle Ojokoh is a Professor in the department of Information Systems of the Federal University of Technology, Akure, Nigeria, with specialty in Artificial Intelligence. She has collaborated with researchers and scientists within Nigeria and from a host of other countries. She has participated and spoken in many international events, including the World Science Forum, South African Science Forum, TWAS Conferences, and Trieste Next Event. She has served as PC member, and Scientific Chair of scientific events, and the Organizing Chair of four editions of the African Symposium on Big Data, Analytics and Machine Intelligence. She has mentored up to 28 Masters students and 15 Ph.D students to graduation. She has over 110 publications to her credit. |
| **Zuber, Maria T.**<br>Maria Zuber is MIT's E. A. Griswold Professor of Geophysics and presidential advisor for science and technology policy. Zuber's research bridges planetary geophysics and the technology of space-based laser and radio systems. Since 1990, she has held leadership roles associated with a dozen scientific experiments or instrumentation on ten NASA missions, most notably serving as Principal Investigator of the Gravity Recovery and Interior Laboratory (GRAIL) mission. Professor Zuber is the first woman to lead a science department at MIT and the first to lead a NASA planetary mission. Zuber holds a B.A. from the University of Pennsylvania and an Sc.M. and Ph.D. from Brown University. |
| **Bess, Lane**<br>Lane Bess is CEO of Deep Instinct and Principal/Founder of Bess Ventures and Advisory. With over 30 years of executive experience in building technology businesses across international regions, achievements in IT security have made Lane a highly trusted advisor to many enterprise CIOs. Lane has been featured in the Wall Street Journal, CNBC, Yahoo! Finance, Forbes, and more. Previously, Lane was COO of Zscaler, a leader in cloud-based internet security, CEO of Palo Alto Networks, as well as EVP of Worldwide Sales and GM of Trend Micro. Lane serves as Board Chairman for Blaize Holdings and Board trustee at Carnegie Mellon, where he earned a Bachelor's of Science. He also holds a Master's Degree from the University of Dayton. Lane is an Astronaut having flown twice on Blue Origin rockets. |

**Mital, Amit**
Amit Mital is Founder and CEO of Verific.ai, a leader in AI governance and compliance. Amit served as Special Assistant to the President, National Security Council at the White House from 2021-2022. He focused on policy for Cyber Security and Emerging Technologies, including Critical Infrastructure security, Telecom policy, Quantum computing and AI. He currently serves on the boards of Cohere Technologies, Deep Instinct and ESentire. Previous roles include CEO of Kernel Labs and EVP/GM and Chief Technology Officer at Symantec Corporation (2013-2015), and Corporate Vice President at Microsoft (1993-2013) responsible for leading strategic product development along with identifying new innovation opportunities. He holds 43 patents, and a M.S. in Engineering from Dartmouth College.

## References

Allahrakha, N. (2025). AI and corruption: Legal liability in algorithmic decision-making. Access to Justice in Eastern Europe, 8(3), 303–326.

Cihon, P., Maas, M. M., & Kemp, L. (2020). Fragmentation and the future: Investigating architectures for international AI governance. Global Policy, 11(5), 545–556.

Cohen, G., Evgeniou, T., Gerke, S., & Minssen, T. (2020). The European artificial intelligence strategy: Implications and challenges for digital health. The Lancet, 2, e379.

de Almeida, P. G. R., dos Santos, C. D., & Farias, J. S. (2021). Artificial intelligence regulation: A framework for governance. Ethics and Information Technology, 23, 505–525.

Djeffal, C., Siewert, M. B., & Wurster, S. (2022). Role of the state and responsibility in governing artificial intelligence: A comparative analysis of AI strategies. Journal of European Public Policy, 29(11), 1799–1821.

Gupta, M., Akiri, C., Aryal, K., Parker, E., & Praharaj, L. (2023). From ChatGPT to ThreatGPT: Impact of generative AI in cybersecurity and privacy. IEEE Access, 11, 80218–80245.

Mügge, D. (2024). EU AI sovereignty: For whom, to what end, and to whose benefit? Journal of European Public Policy, 31(8), 2200–2225.

Nakamura, H., & Kajikawa, Y. (2018). Regulation and innovation: How should small unmanned aerial vehicles be regulated? Technological Forecasting and Social Change, 128, 262–274.

Vaccaro, A. G., Almaatouq, A., & Malone, M. (2024). When combinations of humans and AI are useful: A systematic review and meta-analysis. Nature Human Behavior, 8, 2293–2303.

# Chapter 4
# AI and the Future University: Strategic Leadership, Culture, Diversity and Global Equity

**Yousif Elsamani**

## 1 Opening Vision: A University at a Turning Point

> AI is a concern for all our universities, and it is a concern for university leaders.—**Mosa Moshabela**
>
> We are trying to reconstruct a plane while it is in flight.—**Nick Campbell**
>
> Technology itself is not the goal. The goal is to empower talent and develop human wisdom.—**Qihuang Gong**

Humanity is entering a moment of profound transformation—one where artificial intelligence is reshaping not only research and industry but the very identity and mission of the modern university. The conversations that once centered on whether AI should be used have shifted decisively toward a far more urgent question: What kind of future will universities build with AI, and for whom?

The speakers in this STS *forum* session framed AI not as a technological accessory, but as a defining force requiring deep reflection, strategic leadership, and collective responsibility. The significance of this discussion was underscored by a Nature article that profiled this very session, highlighting how global university leaders see AI as reshaping the foundations of knowledge creation, academic culture, and equity in higher education (*Universities Grapple with the Promise and Perils of AI*, n.d.). The pressures of accelerated technological change are forcing institutions to rethink long-standing assumptions about learning, knowledge, and academic authority. As Nick Campbell observed, universities are being asked to reform themselves while simultaneously guiding society through unprecedented disruption—rebuilding an aircraft mid-flight. Artificial intelligence presents universities with extraordinary opportunities for discovery, creativity, and global collaboration, but it also demands clarity of values, attention to inequality, and humility in governance.

Y. Elsamani (✉)
Institute for Future Initiatives, The University of Tokyo, Tokyo, Japan


H. Komiyama et al. (eds.), *AI, Innovation, and Beyond*,
https://doi.org/10.1007/978-981-95-7020-1_4

> As university leaders, we have to embrace AI … but we will have to figure out how to answer these critical questions together.—**Mosa Moshabela**

This shared responsibility begins with recognizing that universities are no longer passive observers of technological progress. They are expected to act as architects of ethical and social direction, ensuring that AI strengthens rather than undermines democratic knowledge, cultural diversity, and human agency. This means redefining the purposes of education for an age in which generative systems increasingly mediate thought, language, and assessment. It also means building institutional infrastructures and governance models that assert academic autonomy rather than capitulating to commercial monopoly. Above all, it implies affirming that the future of learning belongs to human-centered innovation, not efficiency-driven automation.

Teruo Fujii emphasized that universities carry a dual responsibility: to advance AI research and deployment, while safeguarding society through standards, truth-grounding, and public trust. Universities, he argued, must become "the hub—the global knowledge hub for the future society," not simply adopters of tools created elsewhere. This is a call for leadership that is proactive, principled, and globally engaged.

> As university leaders, we have the responsibility to guide technology rationally and to build constant constraint and innovation mechanisms.—**Qihuang Gong**

This leadership must begin from a philosophical foundation. AI is not merely a technical instrument: it carries assumptions about culture, intelligence, and what it means to be human. If universities limit themselves to transactional uses, grading automation, administrative efficiency, or productivity metrics, they risk abandoning their historical mission as guardians of knowledge, meaning, and culture. Universities must instead model a world in which AI supports, rather than substitutes, human judgment, creativity, wisdom, and empathy.

The urgency of this responsibility is amplified by global competition and geopolitical fragmentation. AI development today is overwhelmingly dominated by a few corporations and a few nations. Brendan Walsh warned that compute inequality and cost barriers threaten to produce a two-tier world: one where privileged institutions access the full power of AI, while others, especially in developing regions, are reduced to passive consumption. The risk is not only technical dependency but epistemic dependency: a future in which knowledge is filtered through systems trained on narrow cultural foundations.

Against this backdrop, universities must exercise their role as global conveners of dialogue, research, and shared governance. This aligns with international scholarship emphasizing responsible and accountable AI integration in institutional strategy (Habbal et al., 2024; Memarian & Doleck, 2023; Papagiannidis et al., 2025) and the importance of shared governance models such as the EU AI Act (Cancela-Outeda, 2024). Emerging studies on scientific discovery further reinforce the transformative potential of AI in research and innovation, while cautioning against closed ecosystems that restrict access to computational capacity and scientific participation (Wang et al., 2023).

Universities therefore stand at an inflection point. AI provides the possibility to reshape learning, expand research horizons, and create more inclusive systems of global knowledge—if stewarded wisely. If not, it may entrench inequity, erode academic autonomy, and displace human judgment with opaque algorithmic authority. The decisions university leaders make now—about collaboration, governance, pedagogy, and openness—will determine which direction prevails.

> The future of AI is here ... and we have to embrace it. But we must be prepared to answer difficult questions—and learn from each other.—**Mosa Moshabela**

This chapter explores how universities can navigate this transformation: how institutional strategy must evolve; how culture and linguistic diversity must be protected; how ethical and environmental responsibilities must guide deployment; and how universities can build a future in which AI strengthens rather than diminishes human purpose. The next section examines the leadership decisions shaping AI implementation across universities globally; revealing how institutional identity, governance, and pedagogical philosophy are being reconstructed in real time.

## 2 Strategic Leadership & Institutional Transformation

> With the emergence of generative AI, we issued guidelines very quickly and urged students to understand what the level of AI itself is and tried to see how they are behaving.—**Teruo Fujii**

> AI is not only for computer science ... it is for everyone.—**Mosa Moshabela**

Across global higher education, artificial intelligence is accelerating a strategic shift in how universities define their missions, build internal capability, and govern technological change. The leaders in this session described an institutional moment in which universities are being forced to move faster and more decisively than at any time in recent history—developing governance models, infrastructure, and educational strategies concurrently, rather than sequentially. The challenge is not merely adopting AI but integrating it into institutional identity while preserving the values that define academic life.

The first wave of actions taken by universities reflects the urgency of this transformation. Responding rapidly to the disruptive appearance of generative AI tools in 2022–2023, the University of Tokyo implemented guidelines to support responsible educational use, framing AI not as a shortcut to answers but as a participant in critical inquiry. Rather than prohibiting usage, the aim was to develop students' ability to interrogate AI, evaluate its reasoning, and incorporate its output transparently into learning processes. This approach reflects a broader movement toward AI fluency and critical literacy—skills increasingly recognized as foundational to learning and research (Schüller, 2022; Wu & Yu, 2024).

> Every member of the university must have an opportunity to learn about AI so that it does not remain a black box.—**Mosa Moshabela**

The shift in strategy is not limited to policy mechanisms; it demands cultural change. Institutions such as the University of Cape Town built multidisciplinary alliances connecting engineering, humanities, education, and health sciences, ultimately launching an AI Institute designed around broad participation rather than narrow specialization. This reflects patterns identified in recent literature emphasizing the importance of interdisciplinary ecosystems in AI innovation, while also showing that the benefits of human–AI collaboration are highly context-dependent and not universally superior to AI alone (Vaccaro et al., 2024) and where knowledge integration accelerates discovery (Wang et al., 2023).

The National University of Singapore echoed this theme, highlighting three pillars of institutional strategy: mission alignment, investment, and governance. Their emphasis on upskilling all members of the university workforce, including 5000 administrative staff who undertook AI training during the pandemic, demonstrates a recognition that digital transformation is not confined to research or teaching, but encompasses the operational fabric of the institution. This aligns with emerging scholarship emphasizing that navigating AI transformation requires robust ethical frameworks, interdisciplinary collaboration, and sustained investment in education and training—not isolated technological adoption (Knopp et al., 2023).

Singapore's focus on community engagement and the co-development of AI solutions with industry partners further illustrates a shift from universities as teaching and research providers to universities as catalysts within national innovation ecosystems. Such partnerships are increasingly essential in a world where computing power, data, and infrastructure are concentrated in a small number of corporate and geopolitical centers (Kitsara, 2022). Universities cannot afford to be passive recipients of commercial platforms; they must negotiate their capacity to act, shape, and govern.

> We need to build constant constraint and innovation mechanisms—using institutional wisdom to guide technology rationally.—**Qihuang Gong**

This leadership requires principled decision-making. Peking University's strategy emphasizes balancing empowerment with educational integrity, innovation with equality, and openness with risk management. Their establishment of a Research Center for AI Security and Governance represents an institutional recognition that responsible AI must be embedded in practice, not aspirational rhetoric. Similar commitments are reflected in recent frameworks addressing AI trust, risk, and security management (Habbal et al., 2024), alignment and safety (Ji et al., 2026; Gabriel, 2020), and responsible governance architectures for AI adoption (Papagiannidis et al., 2025).

ETH Zurich's approach extends this logic through investment in technical infrastructure that enables compliant and trusted use of generative AI. Their establishment of an AI Center linking more than 100 professors, and the development of continuing education programs for industry leaders, signal an evolving understanding of universities as stewards of societal learning and innovation.

> We have the responsibility to build frameworks that allow AI to be used in a trustworthy and compliant manner.—**Günther Dissertori**

This responsibility also extends to the labor and cultural implications of AI. The panelists stressed that efficiency alone cannot be the primary metric of success, as speed-driven deployment risks long-term social and ethical consequences. As Bin Liu noted, leaders must prioritize "human-centric performance over speed, and long-term integrity over short-term efficiency." Human-centered AI scholarship reinforces this principle, arguing that AI should enhance rather than replace human capability and judgment (Shneiderman & Shneiderman, 2022).

The implications for pedagogy are profound. Universities are reimagining the roles of teachers and students: educators are becoming coaches and mentors, facilitating interpretation and reflection rather than transmitting knowledge. AI's capabilities in summarization, translation, and automated feedback reshape learning environments, demanding models built around experiential inquiry and assessment of reasoning rather than output. As Dissertori observed, project-based education, though resource-intensive, may become a primary means of cultivating critical and ethical AI competence.

Universities must exercise global leadership in shaping AI futures. The need for collective action—through alliances like IARU, shared technologies, and open scientific models—was a recurring theme. This aligns with research on global governance concentration and AI inequalities across regions and institutions (Kitsara, 2022). The stakes are high: the future of knowledge production, societal trust, and equitable access to AI will depend not on technological capacity alone, but on the values universities embed within it.

> No one university alone can solve these problems, but collectively we can influence equitable access to AI.—**Brendan Walsh**

Universities therefore stand at a strategic crossroads. They must embrace the innovation imperative while defending human dignity, intellectual integrity, and global inclusivity. Leadership in this context is measured not only by technological capacity, but by moral clarity, the ability to convene diverse communities, and commitment to building AI ecosystems that serve society broadly, not selectively.

These strategic actions illustrate the institutional restructuring already underway, but strategy alone is insufficient. AI raises deeper cultural questions about identity, language, and representation—questions that determine who is included in the future of learning. The next section turns to this challenge by examining how universities can protect linguistic and epistemic diversity in an AI-mediated, English-dominant digital world.

## 3 Cultural & Linguistic Diversity in the Age of AI

> The priority at the University of Cape Town has been to think about how to use AI in the classroom for multilingualism ... so that people feel welcome and included.—**Mosa Moshabela**

> Leading universities must carry out the responsibility of protecting civilizational diversity.—**Qihuang Gong**

As artificial intelligence transforms global education and research, one of the most urgent and overlooked questions is how to protect the diversity of language, culture, and epistemology that constitute the foundation of human knowledge. The dominant architectures of today's large language models (LLMs) are overwhelmingly trained on English-language datasets, and therefore risk narrowing the world's intellectual landscape, reinforcing the dominance of already-privileged cultures. The concern raised repeatedly throughout this STS *forum* discussion is that without deliberate intervention, AI may function as an engine of epistemic inequality—flattening linguistic plurality, marginalizing local knowledge, and standardizing academic expression around a single worldview.

> Everything is done in English … and truth becomes sacrificed.—**Tshilidzi Marwala**

This risk is well documented. Recent analyses of multilingual AI development show that the vast majority of the world's languages are considered low-resource in machine-learning contexts, meaning they lack sufficient digitized training data to be effectively supported by current models (Qin et al., 2025). Emerging literature warns that the erosion of linguistic diversity can not only undermine research and learning, but accelerate the loss of cultural identity, collective memory, and community autonomy (Koc, 2025; Low et al., 2022). AI, without intentional guidance, could quietly become a vehicle for what scholars describe as algorithmic bias and data colonialism, in which dominant datasets serve as proxies for truth (Arora et al., 2023; Ferrara, 2023).

The session speakers framed this not as a technical inconvenience, but as a moral and educational imperative. At the University of Cape Town, multilingual AI translation is being pursued as an instrument of classroom inclusion and cognitive empowerment. Allowing students to think, speak, and write in their home languages, Moshabela argued, alters not only comprehension but dignity. He quoted and echoed Nelson Mandela's principle that "speaking to someone in their mother tongue reaches their heart." The goal is not convenience but belonging—a reminder that language is inseparable from identity.

> We are guardians of linguistic and cultural heritage.—**Bin Liu**

The National University of Singapore positioned its regional ecosystem as a strategic advantage, leveraging Southeast Asia's linguistic richness to drive innovation through the SEA-LION open-source multilingual model. By investing more than $100 million into a publicly accessible platform covering more than 20 languages, Singapore demonstrates a model of inclusive digital infrastructure: AI not as proprietary capital, but as regional commons.

At the same time, Peking University highlighted its multilingual and digital humanities initiative for minority languages and ancient texts, as well as the development of culturally sensitive AI and cross-civilization interpretation platforms. These initiatives position universities as stewards of cultural continuity—responsible not only for educating students, but also for preserving the histories and worldviews embedded in language.

These initiatives respond to growing concerns that large language models may deepen existing digital inequalities; empowering speakers of dominant languages while marginalizing communities whose knowledge systems are underrepresented in training datasets. The consequences reach beyond translation: when AI determines which materials are accessible, which arguments are persuasive, and which knowledge is searchable, it shapes thought.

This challenge is already visible in classrooms. As Gong noted, the release of powerful Chinese models such as DeepSeek transformed usage patterns among students, shifting academic practice and raising questions about fairness, assessment, and educational equity. The uneven distribution of AI literacy and access, whether across countries, languages, or socioeconomic groups, threatens to create a two-tier academic world in which advantage is algorithmically reproduced.

> AI should not be a new source of inequality.—**Qihuang Gong**

Universities therefore face a strategic responsibility: to shape AI ecosystems that expand access rather than restrict it. Large-scale open models such as Switzerland's Apertus, trained on more than 1000 languages with 40% non-English data, signal a move toward transparent and publicly governed AI infrastructure. The creation of such systems suggests a role for universities as counterweights to industry—organizations capable of building AI not primarily for profit, but for global intellectual equity.

> It is a statement: public institutions can make a significant contribution.—**Günther Dissertori**

Yet speakers consistently acknowledged that this problem cannot be solved in isolation. Protecting linguistic diversity requires international collaboration, shared resources, and public infrastructure, not individual experimentation. Universities must form alliances that pool datasets, models, digital archives, and research talent—because the cost of compute, data curation, and model training far exceeds what most institutions or countries can afford independently.

The stakes are clear. Language is not simply a communication tool; it is a way of knowing and thinking. When linguistic and cultural plurality disappear from educational spaces, the diversity of human thought narrows. When AI mediates knowledge production, the systems that shape language also shape epistemology. And when universities relinquish control of this process to private platforms, they surrender intellectual sovereignty.

The path forward demands a reframing of universities as guardians not only of knowledge, but of cultural memory and pluralism. AI must be developed not only to translate languages but to preserve them; not only to optimize learning but to deepen it; not only to accelerate efficiency, but to protect what makes us human. This requires leadership capable of forging shared infrastructures, challenging commercial dominance, and embedding the principles of Ubuntu, solidarity, and global reciprocity at the heart of digital transformation.

The next section turns from cultural responsibility to ethical responsibility—exploring how universities can build governance frameworks that ensure AI remains trustworthy, transparent, and aligned with human values.

## 4 Universities as Stewards of Ethical and Responsible AI

> We have the responsibility to educate current and future leaders … we are building a technical framework that allows the compliant use of generative AI.—**Günther Dissertori**

As artificial intelligence rapidly reshapes the foundations of academic research, teaching, and global collaboration, universities find themselves positioned not only as adopters of innovation but as critical stewards of ethical responsibility. The accelerating pace of AI development, outstripping regulatory capacity and public comprehension, demands institutional leadership capable of guiding society through both profound opportunity and unprecedented risk. Unlike private entities, universities carry a mandate to protect the public interest, preserve human values, and develop frameworks that safeguard collective futures.

> In meetings about AI and AI safety, the dominant voices are often industry leaders rather than academic or public-interest experts.—**Mosa Moshabela**

Speakers highlighted a widening imbalance: whereas commercial stakeholders currently dominate global AI development, governance discussions, and standards-setting processes, academic institutions remain underrepresented. As Moshabela noted, industry voices are often the loudest in shaping public discourse and regulatory negotiation, while universities frequently arrive late or underpowered. The outcome is an AI ecosystem optimized for commercial efficiency rather than human well-being, cultural diversity, or equitable access.

This recognition aligns with emerging scholarship that calls for multi-stakeholder governance models incorporating academic, civic, engineering, and social science perspectives (Papagiannidis et al., 2025; Radanliev, 2025). As AI becomes embedded in critical infrastructure, from healthcare diagnostics to national security and educational assessment, universities must step forward as independent intermediaries capable of defending truth, transparency, and human agency.

A defining aspect of this stewardship is leadership grounded in values rather than speed. Universities face increasing pressure to adopt AI aggressively to increase efficiency, automate academic processes, and enhance competitiveness. However, Bin Liu cautioned against prioritizing short-term gains over long-term societal impact.

> We will weigh human-centric performance over speed … long-term integrity over short-term efficiency.—**Bin Liu**

This stance reflects a growing consensus that responsible AI must be evaluated not only by accuracy and performance metrics but also by social consequences, distributional effects, and ethical safeguards. Efficiency-driven deployment risks generating structural inequality, workforce exclusion, and loss of control over knowledge infrastructure. Liu's framing suggests a paradigm shift toward wisdom-based rather than competition-based innovation.

The panel also emphasized the necessity for institutional governance frameworks. Many universities face intense pressure to adopt AI tools haphazardly, often through individual faculty experimentation or student improvisation, without robust standards for privacy, transparency, or academic integrity. The result is a fragmented policy landscapes and exposure to ethical and legal risks.

> We build institutional infrastructure to guide how to choose tools, how to use them properly, and how to make disclosures.—**Bin Liu**

> Instead of letting students and staff blindly use ChatGPT or whatever tool, we must build compliant frameworks.—**Günther Dissertori**

These statements resonate with institutional governance literature demonstrating that ethical AI systems require integrated policies that combine technical safeguards, clear accountability mechanisms, explainability standards, risk-assessment protocols, and sustainability audits across organizational levels (Qi et al., 2024; Wu et al., 2021). Universities can lead by deploying transparent AI platforms vetted for privacy, fairness, and data security—models that preserve public trust rather than outsourcing critical infrastructure to opaque corporate systems.

Educational transformation is another core responsibility. As AI increasingly mediates knowledge production, universities must redefine the roles of educators and learners. Traditional models positioning professors as sole transmitters of knowledge become incompatible with an environment where AI provides immediate access to computational reasoning, automated synthesis, and personalized tutoring.

> Teachers will have to transform into coaches … training people to use AI critically and ethically.—**Günther Dissertori**

This shift aligns with research demonstrating that, depending on the context and task, human–AI collaboration might outperform either alone when users possess interpretive, evaluative, and ethical competency (Wu et al., 2021). It underscores the need for universal AI and data literacy across disciplines—not confined to computer science, but embedded into humanities, healthcare, social sciences, and administrative development (Chen et al., 2023; Schüller, 2022).

The ethical responsibility of universities also extends beyond campus boundaries. Speakers stressed that universities must support society as a whole (e.g., local businesses, hospitals, governments, and civic organizations) in understanding and responsibly applying AI technologies.

> We have a responsibility to provide continuing education programs tailored to the workforce, and to upskill our own employees.—**Günther Dissertori**

This echoes broader calls for universities to become anchor institutions in national innovation ecosystems, capable of mobilizing knowledge for shared prosperity.

Ethical AI is not merely an academic policy question; it is a prerequisite for economic and social stability.

Energy sustainability was identified as a further critical dimension. The environmental cost of AI—driven by exponential growth in compute-intensive training—is rarely acknowledged, yet increasingly unsustainable (Kaack et al., 2022; Wang et al., 2024; Wu et al., 2021).

> We must take care of the energy issues around AI … and explore alternatives to existing silicon-based systems.—**Teruo Fujii**

This reinforces emerging discussions on aligning AI development with climate goals, recognizing the paradox that AI may simultaneously aid and accelerate environmental degradation (Kaack et al., 2022).

> Perhaps by 2040 an AI agent may win a Nobel Prize … we must think about responsibility across the entire chain of developing AI agents.—**Mosa Moshabela**

Ultimately, the ethical stewardship role of universities can be summarized in three interlinked responsibilities:

- Create governance frameworks that protect integrity, equity, transparency, and inclusion
- Develop educational models that cultivate ethical AI literacy and critical reasoning
- Lead cooperative efforts toward sustainable, fair, open, and accountable AI ecosystems

Without such leadership, universities risk becoming followers in a world shaped primarily by commercial profit and political power rather than social purpose. The next section turns from principles to collective action, examining how universities can work together to address structural inequalities in AI access and governance.

## 5 Toward a Shared Future: Collective Action and the Global Governance of AI

> AI should not be a new source of inequality. It should be a tool to narrow the digital gap and achieve sustainable development.—**Qihuang Gong**

A recurring theme in the discussion was that no single institution or nation can independently shape the trajectory of artificial intelligence. The complexity and scale of the social, ethical, environmental, and geopolitical challenges surrounding AI require coordinated action rather than isolated responses. Speakers emphasized the urgency of building collaborative mechanisms capable of safeguarding equitable access, protecting vulnerable populations, and ensuring that AI development supports collective rather than selective progress.

The digital divide emerged as a central concern. Universities in well-resourced countries are rapidly integrating AI-powered learning platforms, tutoring systems, and administrative automation, while institutions in the Global South continue to struggle with basic infrastructure barriers such as computational capacity and affordability. This asymmetry risks solidifying a two-tiered academic landscape where innovation and opportunity are geographically concentrated.

> The digital divide and affordability barriers will determine who can participate and who can compete in AI.—**Brendan Walsh**

Walsh argued that computing resources such as GPUs and training datasets now represent strategic assets comparable to energy and telecommunications infrastructure. Universities purchasing hundreds of millions of dollars in compute capacity cannot compete with corporations investing tens of billions annually in high-performance AI resources. Academic institutions risk becoming dependent on commercial vendors who control access, pricing, and regulatory boundaries.

Collaboration therefore becomes both a practical necessity and an ethical obligation. Gong stressed the importance of multilateral platforms that unite universities around shared missions; exchange networks, joint research centers, global classrooms, and open-access public resources capable of reducing disparity rather than reinforcing it. Such alliances align with the growing recognition that the future of AI governance must be built cooperatively through cross-border learning, open science, and shared responsibility.

> Promoting inclusive international platforms ... AI must be used to narrow the digital gap.—**Qihuang Gong**

Nick Campbell underscored that the legitimacy of AI adoption depends on transparency and openness, particularly around research that underpins AI systems.

> Leaders must establish principles that guide governance and mitigate risk—even as we reconstruct the plane while it is in flight.—**Nick Campbell**

This metaphor captures the reality that AI governance must evolve dynamically alongside technological change. Regulatory structures that move slowly against rapidly advancing innovation risk becoming obsolete upon implementation. Universities therefore serve as both advisors and challengers, contributing independent expertise in ethical oversight, scientific evidence, and cross-sector dialogue.

Energy consumption was also framed as a shared global responsibility. Fujii argued that sustainable AI development requires new forms of infrastructure innovation, open research on chip alternatives, and international efforts to reduce environmental burden.

> We must address the energy issues around AI and explore alternatives beyond existing silicon-based systems.—**Teruo Fujii**

Environmental sustainability intersects directly with global justice: the communities least responsible for emissions often suffer the greatest consequences. Integrating energy-aware AI development into research agendas aligns with emerging literature supporting planetary stewardship frameworks and climate-aligned innovation (Kaack et al., 2022; Wang et al., 2024).

Audience participants raised concerns regarding academic legitimacy in a world where AI agents generate content indistinguishable from human work. The challenge is not merely regulating AI usage, but re-imagining what learning, scholarship, and assessment mean in a post-generative era. Universities must protect academic integrity while embracing AI's potential to augment learning through more authentic, project-based, and socially embedded models.

As the session concluded, a unifying message emerged: universities cannot afford to approach AI transformation defensively or individually. Leadership requires openness, collaboration, and a willingness to reshape long-standing assumptions about knowledge, equity, and shared responsibility.

The vision articulated by the speakers positions universities not merely as institutions adapting to technological change, but as global stewards shaping the ethical horizon of artificial intelligence. Their collective conviction was clear: the future of AI must be built through shared governance, transparent infrastructure, and international solidarity. The final section synthesizes the lessons of this chapter and outlines practical pathways for global university leadership in advancing ethical, inclusive, human-centered AI.

## 6 Conclusion and Future Directions

Artificial intelligence is reshaping the foundations of higher education, research ecosystems, and global collaboration. Across the discussion, university leaders converged on a shared understanding: AI is not merely a technological disruption, but a societal transformation that demands institutions lead with wisdom, responsibility, and collective purpose.

> The future of AI is here, and as university leaders, we have to embrace it…perhaps we could also try to answer critical questions together and learn from each other.—**Mosa Moshabela**

This collaborative vision aligns with perspectives that position universities as governance intermediaries, bridging research, policy, and public engagement. Rather than limiting their role to teaching and research, universities are now expected to provide ethical stewardship, equitable access, and long-term sustainability. Responsible AI is increasingly integral to institutional mission, not an optional add-on.

Speakers also underscored the urgency of addressing widening gaps in computational resources, AI skills, and infrastructure between regions and institutions. The risk that AI will exacerbate global inequality and concentrate power has been widely documented (Kitsara, 2022). At the same time, projects such as Switzerland's Apertus and Singapore's SEA-LION show how public academic institutions can serve as counterweights to private sector dominance, building multilingual, open, and inclusive AI infrastructure.

A persistent theme centered on values. Universities must defend truth, transparency, and human dignity—principles increasingly challenged by rapid automation, synthetic knowledge, data opacity, and energy-intensive AI infrastructure. This reinforces calls for global AI governance mechanisms and cross-institutional policy alignment, moving beyond fragmented national regulation toward cooperative ethical stewardship.

> Educators will have to transform themselves into coaches…that is the biggest challenge we are facing now.—**Günther Dissertori**

Universities must therefore invest in workforce transformation, continuing education, and community capacity building, extending their leadership beyond campus boundaries.

> Universities should be the global knowledge hub for the future society.—**Teruo Fujii**

Looking ahead, the speakers envisioned universities as global hubs for responsible knowledge ecosystems; places where AI advances scientific discovery, supports linguistic and cultural plurality, and strengthens societal well-being. This vision aligns with a longstanding mission of universities: not only advancing human-centric innovation, but also pursuing fundamental questions about nature, society, and existence, and ensuring that emerging technologies serve these deeper scholarly purposes.

The chapter reveals a turning point. AI challenges universities to protect academic integrity, expand equitable access, build ethical and sustainable systems, and cultivate human wisdom. Real leadership lies not in accelerating adoption for efficiency alone, but in shaping an AI future that elevates society.

The path forward demands cross-border collaboration, open research ecosystems, and culturally grounded AI that honors global diversity. Universities must embrace their dual identity, as innovators and guardians, ensuring that AI becomes not a dividing force, but a catalyst for shared human progress.

**Competing Interests** No Competing interests.

## Session Panelists

**Dr. Teruo Fujii** is the 31st President of the University of Tokyo. Prior to taking the President's office in April 2021, he was Executive Vice President in charge of finance and external relations. He served as Director General of the Institute of Industrial Sciences (IIS) from 2015 to 2018. He received his PhD in engineering from UTokyo in 1993. After working at RIKEN, he joined IIS in 1999 as an associate professor and became a full professor in 2007. He served as an Executive Member of the Council for Science, Technology and Innovation from 2021 to 2024. He was also Co-director of LIMMS-CNRS/IIS from 2007 to 2014, and the President of Chemical and Biological Microsystems Society from 2017 to 2019. His research specializes in applied microfluidics systems and underwater technology.

| |
|---|
| **Professor Mosa Moshabela** is the Vice-Chancellor and Principal of the University of Cape Town. He is a member of Council and Chair of Senate. He is the Chairperson of the Governing Board at the National Research Foundation (NRF), a former Member of the Board at the South African Medical Research Council (SAMRC), and former Chairperson of the Standing Committee on Health in the Academy of Science of South Africa (ASSAf). Globally, he is a member of the international advisory board for the Lancet Healthy Longevity, Lancet commission on synergies between Health Promotion, Universal Healthcare Access and Global Health Security, and the commission of the US National Academies for Science, Engineering and Medicine (NASEM) on the Global Roadmap to Healthy Longevity. |
| President of Peking University, **Prof. Gong Qihuang** is an academician of the Chinese Academy of Sciences and a fellow of the World Academy of Sciences. He also serves as a standing committee member of the China Association for Science and Technology, and Vice Chairman of the Beijing Association for Science and Technology. Professor Gong received his PhD from PKU and has long been engaged in cutting-edge research on nonlinear optics and spatiotemporal small-scale optics. An elected Fellow of the Institute of Physics (UK), the Optical Society of America, and the Society of Photo-Optical Instrumentation Engineers, he was previously the President of the Chinese Optical Society, Vice President of the International Commission for Optics, and Vice President of the Chinese Physics Society. |
| **Professor Bin Liu** is Tan Chin Tuan Centennial Professor at the National University of Singapore (NUS). Bin graduated with bachelor's degree from Nanjing University and a Ph.D. in Chemistry from NUS. She had postdoctoral training at the University of California, Santa Barbara before joining NUS in late 2005. Bin has been well-recognized for her contributions to polymer chemistry and organic nanomaterials for energy and biomedical applications. She is an international member of the US National Academy of Engineering. Since 2019, she has served as the Deputy Editor to launch and develop ACS Materials Letters, a flagship materials journal of the American Chemical Society. |
| **Günther Dissertori** has been Rector of ETH Zurich since Feb 2022, and Professor for Particle Physics at ETH Zurich since 2001. He was Head of the Institute for Particle Physics and Astrophysics, as well as Director of Studies of the Physics Department for several years. Before joining ETH Zurich, he was research scientist and fellow at CERN, the European laboratory for Particle Physics in Geneva, Switzerland. He obtained his PhD in 1997 from the University of Innsbruck, Austria. His research focus is on particle physics at colliders, in particular using the CMS experiment at CERN's Large Hadron Collider. With his group he had a leading involvement in the discovery of the Higgs boson in 2012. In addition, his group has developed novel Positron Emission Tomography scanners. |
| **Tshilidzi Marwala** is the Rector of the United Nations University and a UN Under-Secretary-General. Formerly the Vice-Chancellor of the University of Johannesburg, he is a distinguished South African scholar, academic administrator, and author of over 25 books on leadership and technology in the 21st century. |
| **Nick Campbell** is the Vice President for Academic Affairs at Springer Nature, where he has held various leadership roles since joining in 2001. His career includes experience as an editor for journals like Nature Reviews Genetics and as the Managing Editor of Nature, leading a major digital and print relaunch. He has a background in genetics, holding a PhD and postdoctoral research in the field. |

**Brendan M. Walsh** leads Yale University's global strategy and oversees a team of 49 staff members. He advises university leadership on international issues that affect Yale's mission and manages an annual budget of approximately $12 million. Previously, he served as Director of International Affairs at Stanford University from 2011 to 2024. Before that, he was a Senior Officer at the U.S. Department of State from 2006 to 2011, a Senior Researcher at New York University from 2004 to 2006, and Assistant Director for International Students and Scholars at Loyola Marymount University from 2001 to 2004. He holds a doctorate from New York University and a B.A. in History, with a minor in German, from the College of the Holy Cross. His honors include the Distinguished Service Award from the Scholars at Risk Network, the Superior Honor Award from the U.S. Department of State, the Outstanding Doctoral Dissertation Award from New York University, and a Fulbright Teaching Fellowship.

## References

Arora, A., Barrett, M., Lee, E., Oborn, E., & Prince, K. (2023). Risk and the future of AI: Algorithmic bias, data colonialism, and marginalization. *Information and Organization*, *33*(3), 100478. https://doi.org/10.1016/j.infoandorg.2023.100478

Cancela-Outeda, C. (2024). The EU's AI act: A framework for collaborative governance. *Internet of Things*, *27*, 101291. https://doi.org/10.1016/j.iot.2024.101291

Chen, R. J., Wang, J. J., Williamson, D. F. K., Chen, T. Y., Lipkova, J., Lu, M. Y., Sahai, S., & Mahmood, F. (2023). Algorithmic fairness in artificial intelligence for medicine and healthcare. *Nature Biomedical Engineering*, *7*(6), 719–742. https://doi.org/10.1038/s41551-023-01056-8

Ferrara, E. (2023). Fairness and Bias in Artificial Intelligence: A Brief Survey of Sources, Impacts, and Mitigation Strategies. *SSRN Electronic Journal*. https://doi.org/10.2139/ssrn.4615421

Gabriel, I. (2020). Artificial Intelligence, Values, and Alignment. *Minds and Machines*, *30*(3), 411–437. https://doi.org/10.1007/s11023-020-09539-2

Habbal, A., Ali, M. K., & Abuzaraida, M. A. (2024). Artificial Intelligence Trust, Risk and Security Management (AI TRiSM): Frameworks, applications, challenges and future research directions. *Expert Systems with Applications*, *240*, 122442. https://doi.org/10.1016/j.eswa.2023.122442

Ji, J., Qiu, T., Chen, B., Zhou, J., Zhang, B., Hong, D., Lou, H., Wang, K., Duan, Y., He, Z., Vierling, L., Zhang, Z., Zeng, F., Dai, J., Pan, X., Xu, H., O'Gara, A., Ng, K., Tse, B., … Gao, W. (2026). AI Alignment: A Contemporary Survey. *ACM Computing Surveys*, *58*(5), 1–38. https://doi.org/10.1145/3770749

Kaack, L. H., Donti, P. L., Strubell, E., Kamiya, G., Creutzig, F., & Rolnick, D. (2022). Aligning artificial intelligence with climate change mitigation. *Nature Climate Change*, *12*(6), 518–527. https://doi.org/10.1038/s41558-022-01377-7

Kitsara, I. (2022). Artificial Intelligence and the Digital Divide: From an Innovation Perspective. In A. Bounfour (Ed.), *Platforms and Artificial Intelligence: The Next Generation of Competences* (pp. 245–265). Springer International Publishing. https://doi.org/10.1007/978-3-030-90192-9_12

Knopp, M. I., Warm, E. J., Weber, D., Kelleher, M., Kinnear, B., Schumacher, D. J., Santen, S. A., Mendonça, E., & Turner, L. (2023). AI-Enabled Medical Education: Threads of Change, Promising Futures, and Risky Realities Across Four Potential Future Worlds. *JMIR Medical Education*, *9*(1), e50373. https://doi.org/10.2196/50373

Koc, V. (2025). Generative AI and Large Language Models in Language Preservation: Opportunities and Challenges (Version 2). arXiv. https://doi.org/10.48550/ARXIV.2501.11496

Low, D. S., Mcneill, I., & Day, M. J. (2022). Endangered Languages: A Sociocognitive Approach to Language Death, Identity Loss, and Preservation in the Age of Artificial Intelligence. *Sustainable Multilingualism*, *21*(1), 1–25. https://doi.org/10.2478/sm-2022-0011

Memarian, B., & Doleck, T. (2023). Fairness, Accountability, Transparency, and Ethics (FATE) in Artificial Intelligence (AI) and higher education: A systematic review. *Computers and Education: Artificial Intelligence*, *5*, 100152. https://doi.org/10.1016/j.caeai.2023.100152

Papagiannidis, E., Mikalef, P., & Conboy, K. (2025). Responsible artificial intelligence governance: A review and research framework. *The Journal of Strategic Information Systems*, *34*(2), 101885. https://doi.org/10.1016/j.jsis.2024.101885

Qi, X., Huang, Y., Zeng, Y., Debenedetti, E., Geiping, J., He, L., Huang, K., Madhushani, U., Sehwag, V., Shi, W., Wei, B., Xie, T., Chen, D., Chen, P.-Y., Ding, J., Jia, R., Ma, J., Narayanan, A., Su, W. J., … Mittal, P. (2024). *AI Risk Management Should Incorporate Both Safety and Security* (Version 1). arXiv. https://doi.org/10.48550/ARXIV.2405.19524

Qin, L., Chen, Q., Zhou, Y., Chen, Z., Li, Y., Liao, L., Li, M., Che, W., & Yu, P. S. (2025). A survey of multilingual large language models. *Patterns*, *6*(1), 101118. https://doi.org/10.1016/j.patter.2024.101118

Radanliev, P. (2025). AI Ethics: Integrating Transparency, Fairness, and Privacy in AI Development. *Applied Artificial Intelligence*, *39*(1), 2463722. https://doi.org/10.1080/08839514.2025.2463722

Schüller, K. (2022). Data and AI literacy for everyone. *Statistical Journal of the IAOS*, *38*(2), 477–490. https://doi.org/10.3233/SJI-220941

Shneiderman, B., & Shneiderman, B. (2022). *Human-Centered AI*. Oxford University Press.

*Universities grapple with the promise and perils of AI*. (n.d.). Retrieved March 30, 2026, from https://www.nature.com/articles/d42473-025-00330-4

Vaccaro, M., Almaatouq, A., & Malone, T. (2024). When combinations of humans and AI are useful: A systematic review and meta-analysis. *Nature Human Behaviour*, *8*(12), 2293–2303. https://doi.org/10.1038/s41562-024-02024-1

Wang, H., Fu, T., Du, Y., Gao, W., Huang, K., Liu, Z., Chandak, P., Liu, S., Van Katwyk, P., Deac, A., Anandkumar, A., Bergen, K., Gomes, C. P., Ho, S., Kohli, P., Lasenby, J., Leskovec, J., Liu, T.-Y., Manrai, A., … Zitnik, M. (2023). Scientific discovery in the age of artificial intelligence. *Nature*, *620*(7972), 47–60. https://doi.org/10.1038/s41586-023-06221-2

Wang, Q., Li, Y., & Li, R. (2024). Ecological footprints, carbon emissions, and energy transitions: The impact of artificial intelligence (AI). *Humanities and Social Sciences Communications*, *11*(1), 1043. https://doi.org/10.1057/s41599-024-03520-5

Wu, C.-J., Raghavendra, R., Gupta, U., Acun, B., Ardalani, N., Maeng, K., Chang, G., Behram, F. A., Huang, J., Bai, C., Gschwind, M., Gupta, A., Ott, M., Melnikov, A., Candido, S., Brooks, D., Chauhan, G., Lee, B., Lee, H.-H. S., …, Hazelwood, K. (2021). *Sustainable AI: Environmental Implications, Challenges and Opportunities* (Version 2). arXiv. https://doi.org/10.48550/ARXIV.2111.00364

Wu, R., & Yu, Z. (2024). Do AI chatbots improve students learning outcomes? Evidence from a meta-analysis. *British Journal of Educational Technology*, *55*(1), 10–33. https://doi.org/10.1111/bjet.13334

# Part II
# Sustainability and Society

# Chapter 5
# Path to Sustainability toward a Zero-Carbon Society

**Yuya Kajikawa**

## 1 At a Pivotal Moment toward a Zero-Carbon Society

> We meet at a pivotal moment in shaping a sustainable, zero-carbon future. Nearly 90% of the world's economies have set climate commitments. This signals unprecedented momentum, yet the challenge is complex.—**Munir M. Eldesouki**

Climate change is not a future issue but the current one we face. Extreme weather has already impacted economic activities and had financial impacts. Losses from major storms, floods, and heatwaves are linked to reduced and more unstable earnings and cash flows of companies (Huang et al., 2018). They are being forced to develop a business continuity plan to take extreme weather and resulting supply chain disturbance into account. Companies in regions with harsher weather tend to keep higher cash reserves to enhance financial flexibility and resilience against climate risks. Those firms also tend to have less short-term debt but more long-term debt, and to be less likely to distribute cash dividends.

Climate change also has a significant impact on agriculture and food security, and the effects vary by region and crop (Anderson et al., 2020). To meet the demands of a growing population, agricultural adaptation is urgently needed, requiring changes in cultivation methods, crop breeding, and a reevaluation of dietary habits. Advances in fields such as genomics can contribute to these efforts, but their implementation requires investment and time. Since there are limits to adaptation, reducing the use of fossil fuels is essential in the long term.

Hot environments and the associated heat stress increase mortality and morbidity and have adverse effects on pregnancy and mental health (Ebi et al., 2021). Furthermore, they impair work capacity and cognitive function, affecting productivity and occupational health risks. Approximately half of the world's population and more than 1 billion workers are exposed to high temperatures, and many of them

Y. Kajikawa (✉)
Institute for Future Initiatives, The University of Tokyo, Tokyo, Japan



H. Komiyama et al. (eds.), *AI, Innovation, and Beyond*,
https://doi.org/10.1007/978-981-95-7020-1_5

suffer health impacts. Climate change is making extreme heat the new normal, and health risks are projected to increase further in the future. In tropical regions, in particular, the likelihood of reaching physiological limits is rising. Urbanization and an aging population also play a role, and in urban areas, artificial heat contributes to rising temperatures.

Imminent threats are apparent as illustrated above. We are at the cornerstone facing climate change and tipping points to mitigate it. However, it seems like we're going back and forth between a zero-carbon society and backlash against it. Achieving a zero-carbon society is at the top of the agenda in nations around the globe. Maximum use of renewable energies with improved energy efficiency and productivity constitutes a key pathway to achieving both a sustainable society and economic growth. On the other hand, conflicts in Europe and the Middle East have elevated national and regional imperatives for energy security and resiliency, bringing new scrutiny to energy sources. In the face of these geopolitical factors, the pressing threats to sustainability appear to have lost their priority, having been supplanted by energy and economic security, while there are few reliable measures of security (Axon & Darton, 2021). For instance, securing pipelines and sea lines for nations is more important than safeguarding the global commons. In addition, the cost of energy in Europe is about two to three times higher than in the US, exerting competitive pressures across industrial sectors, impacting industrial base capacity, disrupting critical supply chains, and weakening enterprise resiliency. Transition to a zero-carbon society is a complex challenge. Because of its complexity, designing a solution for the transition to a zero-carbon society is not a rudimentary task. And because of its complexity, it does not strike a chord with people's emotions in the same way that discourse on national security does.

> Our foremost challenge is a global systems challenge. Sustainability writ large is a multidimensional problem that operates at the intersection of energy production and consumption: how we power and light our world, how we move things, how we build things, how we manage physical and virtual environments, how we extract and utilize critical natural resources, how we develop, produce, and recycle natural and new synthetic materials—and the list goes on.—**Deborah L. Wince-Smith**

Achieving a zero-carbon society through sustainability requires coordinated action across the entire product lifecycle, using integrated strategies and advanced tools to optimize global production networks while engaging with complex systems spanning cities, rural areas, agricultural lands, forests, oceans, and beyond. This "system of systems" challenge unfolds within a fragmented yet interconnected global economy shaped by material and monetary flows, diverse stakeholders, and rapid technological disruption.

> The path to a zero-carbon society is an innovation revolution. Success requires us to address two challenges simultaneously: to harmonize global action toward ambitious climate goals, and to manage the resource implications of emerging technologies.—**Munir M. Eldesouki**

Climate change is the most vital threat to humankind, but the challenges to mitigate are also business opportunities and economic development. As empirically shown by historical data (Shahbaz et al., 2020), energy consumption and financial development lead to environmental degradation. But research and development (R&D) contributes to emissions reduction, and leveraging finance and R&D is crucial. Emerging technologies for mitigating climate change have been the essential target of both public and private investments.

> We have already established a $1.5 billion Sustainability Venture Capital Fund a few years ago, and we have already invested in more than 40 companies globally and in Saudi Arabia that are focused on technologies such as carbon capture, lower-carbon fuels, new energies, advanced materials, and other disruptive technologies.—**Ahmad O. Al-Khowaiter**

The session mainly focuses on three technological domains, namely, carbon capture and storage (CCS), nuclear power, and renewable energies.

## 2 Fossil Fuels and Carbon Capture and Storage

> We're investing in carbon capture and storage, and we're building one of the world's largest carbon capture and storage hubs on the east coast of Saudi Arabia. We're also investing in new carbon capture technologies and advanced capture materials, and we're implementing AI-driven efficiency improvements to reduce overall emissions across the value chain.—**Ahmad O. Al-Khowaiter**

Carbon Capture and Storage (CCS) is a technology that captures carbon dioxide ($CO_2$) emitted from power plants and factories before it is released into the atmosphere, compresses and transports it, and then stores it deep underground on land and in the ocean. It primarily targets $CO_2$ generated by the combustion of fossil fuels and industrial processes (such as steel, cement, and chemicals), with the aim of reducing emissions. The captured $CO_2$ is transported via pipelines or ships and injected into geological formations such as depleted oil and gas fields or deep aquifers. Global saline aquifers and depleted oil and gas reservoirs have sufficient capacity to store up to two centuries' worth of human-generated $CO_2$ emissions (Lau et al., 2021). This allows the $CO_2$ to be sequestered for long periods, preventing its accumulation in the atmosphere.

CCS is considered a viable option because it is highly compatible with current fossil fuel-based energy systems and can be retrofitted into existing facilities such as thermal power plants. CCS is regarded as a critical complementary technology in sectors where emissions reductions remain challenging, even as the transition to renewable energy progresses. In particular, in industries such as cement and steel, $CO_2$ is generated directly from the manufacturing processes themselves, meaning that emissions reductions through energy transition alone have their limits. Furthermore, since completely phasing out existing fossil fuel infrastructure takes time and is costly, CCS is also important as a means of reducing emissions during the transition period. Furthermore, technologies such as BECCS (Bioenergy with Carbon Capture and Storage), when combined with bioenergy, offer the potential to achieve negative emissions, making CCS an indispensable technology for meeting the goals of the Paris Agreement.

CCS primarily consists of three stages, i.e., capture, transport, and storage. There are several types of capture methods in particular. The most common are post-combustion capture, pre-combustion capture, and oxy-fuel combustion. Post-combustion capture is a method that separates $CO_2$ from flue gas and is relatively easy to apply to existing facilities. Pre-combustion capture is highly efficient because it separates $CO_2$ through a gasification process, but it is intended for new facilities. Oxy-fuel combustion is a method that uses pure oxygen for combustion to produce high-concentration $CO_2$. The captured $CO_2$ is transported via pipelines or ships and injected into stable underground formations. By combining these technologies, it is possible to address a wide range of emission sources. In addition to sequestration, captured $CO_2$ can be utilized as a carbon source for the production of chemicals, concrete, etc., which is called carbon capture and utilization (CCU).

> Flagship projects highlight this ambition: from the world's largest carbon capture hub led by Aramco, to the green hydrogen and ammonia facility at NEOM, to one of the world's largest operational battery storage systems in Bisha. These efforts demonstrate how a nation can align its strengths with its responsibilities.—**Munir M. Eldesouki**

A prime example of a CCS project is the Sleipner project in Norway. Launched in 1996, it stores $CO_2$ separated during natural gas production in subsea formations in the North Sea. Additionally, at the Boundary Dam power plant in Canada, $CO_2$ is captured from coal-fired power generation. Furthermore, the Gorgon project in Australia is another large-scale CCS initiative. In recent years, there has been a rise in cluster-type projects—primarily in Europe—that share $CO_2$ transport and storage infrastructure, with multiple companies collaborating to reduce costs and improve efficiency. Flagship CCS projects by Aramco, in collaboration with King Abdullah University of Science and Technology (KAUST), are also challenging direct air capture (DAC) and green hydrogen, which will lead to negative emissions. These examples serve as important demonstrations for the practical application and expansion of CCS. These efforts demonstrate how a nation can align its strengths with its responsibilities, as quoted above.

Investment in CCS has been expanding rapidly in recent years. This growth is driven by national governments' decarbonization policies, the introduction of carbon pricing, and the establishment of subsidy programs. Furthermore, major energy companies and financial institutions have identified CCS as a key investment area and are actively involved in project development and infrastructure development. Additionally, the growth of ESG investing has heightened the recognition of CCS as a technology for reducing emissions. On the other hand, high costs and uncertainty regarding return on investment remain challenges, making continued policy support essential.

There are several challenges associated with CCS. First is the high cost; in particular, the $CO_2$ capture process consumes a significant amount of energy, making it difficult to ensure economic viability. Second are concerns regarding the safety of storage; social acceptance is not yet sufficient regarding long-term leakage risks and environmental impacts. Third, the transportation and storage infrastructure is insufficient, meaning large-scale deployment will take time. There are also institutional challenges, such as regulations and the allocation of responsibility. Furthermore, there is criticism that CCS merely prolongs the use of fossil fuels. To overcome these challenges, technological innovation, policy support, and the formation of social consensus are all essential.

# 3 Advanced Nuclear Power Generation

> Nuclear technology is an important option for countries that want to shift away from coal and gas baseload and cut emissions. So today, I tell you, for the first time in our 60-year history, ADB is prepared to support nuclear power in our operations. Momentum is growing as a new generation of large reactors and small modular reactors emerge. These offer flexible and scalable solutions, especially for countries with limited grid capacity or space for conventional plants.—**Masato Kanda**

Nuclear power generation is a method of generating electricity by utilizing the thermal energy produced by the fission of nuclear fuel, such as uranium, to convert water into steam and drive turbines. Because it has almost no $CO_2$ emissions during power generation, it is considered to play a crucial role in addressing climate change. Among various types of nuclear reactors, including light-water reactors (pressurized water reactors and boiling water reactors), heavy-water reactors, and fast breeder reactors, light-water reactors are currently the mainstream technology. R&D of next-generation technologies that offer enhanced safety and efficiency has been advancing in recent years. Small Modular Reactors (SMRs) are especially attracting attention, as they are expected to reduce costs and improve safety by being manufactured in factories and assembled on-site. In the United States, demonstration projects for small modular reactors are actively underway, marking an important step toward commercialization. In addition, R&D on nuclear fusion is attracting interest, which includes magnetic confinement methods (called tokamak and stellarator) and inertial confinement methods (laser fusion). In the ITER (International Thermonuclear Experimental Reactor) project, more than 35 members are joining.

> Globally, nuclear power combined with renewables is recognized as an essential pillar of the low-carbon mix, as highlighted at COP 27 and COP 28. In Europe, and particularly in France, we have ambitious goals: reduce emissions by 55% by 2030 and reach carbon neutrality by 2050. But the initial strategy favoring renewables at the expense of nuclear has weakened our energy security and competitiveness.—**Claude Imauven**

Nuclear power generation is expected to balance economic feasibility, environmental sustainability, and energy security. The Inflation Reduction Act (IRA) of 2022 marks a historic commitment by the United States to reducing climate-related investment risks. Even in the early 2000s, the U.S. sought to stimulate a nuclear power renaissance through targeted legislative derisking, but it resulted in disappointing results. Across several southern states, these policies encouraged some nuclear investment yet ultimately led to tens of billions of dollars in wasted capital with minimal expansion of carbon-free electricity generation. The lessons encourage us to be careful in regulatory, scale, temporal, and cultural risks. These overlapping dimensions suggest that the IRA's focus on financial derisking alone is insufficient to drive the systemic changes needed for a low-carbon transition by nuclear power (Welton & Harrison, 2025).

Another concern is the dilemma of global dissemination of nuclear power technologies and the non-proliferation principle. The continuous operation of reactors for decades creates opportunities for non-state actors to divert nuclear material that goes undetected for extended periods. Strong nuclear safeguards are needed, and the deployment of Safeguards-by-Design principles in the design and construction phases will be an effective and efficient way, which was enforced decades after the first nuclear power plants started operation (Grape et al., 2014).

## 4 Renewable Energy

> Saudi Arabia has embraced this dual approach through the Circular Carbon Economy framework. It is a pragmatic and holistic pathway, integrating renewable deployment with carbon management and new clean energy frontiers.—**Munir M. Eldesouki**

Renewable energy refers to the generation of electricity, heat, and fuel using energy sources that are virtually inexhaustible in nature. Typical examples include solar, wind, hydro, geothermal, and biomass energy. A key characteristic is that, unlike fossil fuels, these sources are regarded as not finite and do not pose a risk of depletion. Renewable energy is a method of directly obtaining energy by utilizing natural cycles and phenomena; for example, solar power converts light energy into electricity, while wind power converts the kinetic energy of the wind into rotational motion to generate electricity. In recent years, technological innovations have led to improvements in conversion efficiency and cost reductions, driving adoption in various environments, including urban and rural areas as well as offshore locations.

Currently, R&D for 100% renewable systems is an active research topic. Although the regions where these technologies can be implemented vary depending on their physical mechanisms and installation conditions, they are attracting global attention as a sustainable means of power supply. Furthermore, renewable energy functions as a distributed power source, contributing to the flexibility of the power grid and enhancing resilience during disasters. However, because of the temporal and seasonal fluctuation of renewable energy sources, energy storage technologies are necessary to stabilize electric power networks. Thus, an integrated design of renewables, batteries, and hydrogen is needed for cost-efficient design of the electricity system (Dawood et al., 2020; Mathiesen et al., 2015).

> Looking at the international situation in recent years, the energy security aspect has taken on greater importance. Such circumstances highlight the importance of both renewable energy and nuclear power, both of which provide zero-carbon electricity.—**Akihiko Kato**

Energy resources and their uneven distribution have increased international tensions and become one of the major factors in conflicts and wars. Domestic production and distribution of electricity and energy lessen such tension and contribute to global security. Unlike fossil fuels, which are geographically concentrated and often associated with resource scarcity, renewable energy resources such as solar and wind are more widely distributed. This shift has the potential to reduce energy dependency, weaken the geopolitical leverage of major fossil fuel exporters, and enhance energy security for importing countries. As quoted above, renewable energy and nuclear power are placed side by side in the context of energy security, but they still have remaining concerns about it in different ways.

In the case of nuclear power, there is a proliferation risk and governance to prevent military applications. In the case of renewables, the emerging geopolitical risk is critical materials (e.g., rare earth elements, lithium, cobalt) required especially for batteries (Vakulchuk et al., 2020). In addition, control over technology, intellectual property, and manufacturing capacity, particularly in solar panels, batteries, and grid infrastructure, becomes increasingly important in shaping global influence. The energy transition is not geopolitically neutral; rather, it creates a new landscape characterized by shifting dependencies, emerging strategic resources, and changing patterns of power. However, regardless of these remaining concerns, renewable energy can lead to both decentralization and regionalization of energy systems, reducing the likelihood of resource conflicts while increasing the importance of regional cooperation and cross-border electricity and energy trade via grid and hydrogen.

## 5 Global Partnership and Harmonization of Global Policy

> The issue of climate change is a challenge that all of humanity must address together. "One goal, various pathways" is a keyword of Japan's climate change policy.—**Takehiko Matsuo**

As illustrated at the beginning of this chapter, we are at a pivotal moment toward a zero-carbon society. Climate change is not a future issue but the current one we face. However, the movement to a zero-carbon society is disrupted under the rising international tensions and the growing priority placed on geopolitics, although it is not a new issue. War, conflict, crime, and corruption have been major threats to a sustainability transition in myriad ways (Kates & Parris, 2003). However, the transition to renewable energy systems enables a decentralized energy supply system and will lessen international tensions. In addition to the context of decentralization and energy security, actions toward a zero-carbon society serve as an opportunity for international collaborations and global partnerships. We can see many flagships of global collaborations in the session, as quoted below.

> Thailand recognizes that achieving net zero requires global solidarity. We are working closely with ASEAN countries to build regional resilience and collaborating with Japan and the EU on hydrogen technologies, the United States on carbon capture and storage, and China on renewable energy innovation.—**Supachai Pathumnakul**

> International partnerships amplify these efforts. For example, with UC Berkeley, we are pioneering research in nanomaterials essential for carbon capture, water harvesting, and clean energy. This reflects our conviction that no nation can achieve a zero-carbon society alone. Progress depends on collaboration across borders and disciplines.—**Munir M. Eldesouki**

These partnerships are not the same as traditional foreign aid from the Global North to the Global South and Official Development Assistance. As discussed in the science diplomacy chapter in this book, it centers economic agency with economic flow of resources rather than epistemic contribution, including addressing shared challenges and advancing knowledge. This prioritization reflects material realities for nations struggling with poverty, infrastructure deficits, and limited fiscal capacity. Science diplomacy, in this context, becomes meaningful only insofar as it facilitates technology transfer, enables market access, builds domestic capacity, and ensures equitable participation in economic value creation processes.

> When determining how to address climate goals and long-term lower emissions, it is imperative that we apply a full lifecycle, technology-inclusive approach to drive policy. If we do not, we face the risk of creating more emissions, not less.—**Ahmad O. Al-Khowaiter**

The transition toward a decarbonized society requires the deployment of a wide range of policy instruments and financial mechanisms that operate across multiple levels of governance and sectors of the economy. These instruments can be broadly categorized into regulatory approaches, market-based mechanisms, fiscal policies, financial innovations, and institutional frameworks, all of which interact to shape incentives for reducing greenhouse gas emissions and promoting low-carbon technologies.

A central pillar of climate policy is the implementation of carbon pricing mechanisms, which internalize the social cost of carbon emissions but need public trust (Klenert et al., 2018). Carbon taxes directly impose a fee on the carbon content of fossil fuels, thereby encouraging emissions reductions through price signals. Maintaining an effective carbon price and successfully implementing a carbon tax depend not only on economic efficiency but also on political and social acceptability. A central requirement is the transparent and equitable use of carbon pricing revenues. The authors argue that redistributing revenues through mechanisms such as lump-sum rebates, reductions in distortionary taxes, or targeted support for low-income households can offset regressive impacts and increase public support. Without such measures, carbon taxes risk being perceived as unfair, undermining their durability. Institutional credibility and policy stability are also critical. Governments must establish clear, long-term policy frameworks that provide predictable carbon price signals to guide investment in low-carbon technologies. Frequent policy reversals or unclear commitments can weaken investor confidence and reduce the effectiveness of pricing mechanisms. Furthermore, effective communication strategies are necessary to inform the public about the environmental and economic benefits of carbon pricing, as well as how revenues are used.

Emissions trading systems (ETS), or cap-and-trade schemes, which set an overall emissions cap and allow firms to trade allowances, are another instrument to create a market for carbon reduction. Complementing these systems are carbon offset mechanisms and carbon credit markets, where verified emissions reductions or removals—generated through projects such as reforestation, renewable energy deployment, or carbon capture—can be traded to meet compliance or voluntary targets. These include both compliance markets under regulatory regimes and voluntary carbon markets driven by corporate net-zero commitments.

In addition to pricing mechanisms, governments employ a variety of regulatory and command-and-control policies to create carbon markets and to function as an ETS. These include emissions performance standards for power plants and industrial facilities, fuel economy standards for vehicles, renewable portfolio standards (RPS) that mandate a minimum share of electricity from renewable sources, and phase-out regulations targeting coal-fired power generation or internal combustion engine vehicles. Building codes and energy efficiency standards for appliances and infrastructure also play a critical role in reducing energy demand and associated emissions. International mechanisms and cooperative frameworks further support decarbonization efforts. Paris Agreement's nationally determined contributions (NDCs) are an essential pledge. However, policy instruments preventing carbon leakage are also necessary.

A Carbon Border Adjustment Mechanism (CBAM) is a policy instrument designed to equalize the carbon cost of domestic and imported goods. It applies a carbon price to imports based on their embedded greenhouse gas emissions, ensuring that foreign producers face similar regulatory burdens as domestic firms operating under carbon pricing regimes such as carbon taxes or emissions trading systems. Typically, importers are required to purchase certificates reflecting the carbon content of their products, aligned with the prevailing domestic carbon price. CBAMs are primarily applied to emissions-intensive and trade-exposed sectors such as steel, cement, aluminum, fertilizers, and electricity. By incorporating carbon costs into international trade, CBAM aims to prevent distortions caused by uneven climate policies across countries. It effectively extends domestic climate policy beyond national borders, transforming climate regulation into a trade-related instrument. While conceptually straightforward, its implementation requires robust methodologies for calculating embedded emissions, verification systems, and alignment with international trade rules, particularly those under the World Trade Organization (WTO).

However, if the introduction of CBAM is limited to certain regions like the EU, its effectiveness to prevent carbon leakage is also limited, and thus, harmonization of global policy is desirable. The global adoption of CBAM-like policies requires enhanced international cooperation and institutional alignment. One key approach is to promote convergence in carbon pricing systems across countries, reducing the need for border adjustments over time. Multilateral forums such as the G20, OECD, and WTO can facilitate dialogue and establish common principles for carbon pricing and trade-related climate measures. Another important step is the development of standardized methodologies for measuring emissions, enabling mutual recognition and reducing compliance costs across borders. Financial and technical assistance to developing countries is essential to ensure inclusiveness and prevent CBAM from exacerbating global inequalities. Linking emissions trading systems across regions could also foster cooperation and reduce fragmentation. Additionally, incorporating CBAM principles into international climate agreements, such as under the Paris Agreement framework, could strengthen legitimacy and coordination. Ultimately, a cooperative approach that balances environmental ambition, economic fairness, geopolitical considerations, and just transition is necessary to achieve widespread adoption and effectiveness of CBAM worldwide.

## 6 Responsible Innovation and Investments: New Role of Finance and Investors

> The way forward is a dual innovation strategy. First, we must scale proven, cost-effective solutions: solar, wind, nuclear, and efficiency. At the same time, we must invest in research and innovation to unlock the 40% of emissions reductions that depend on technologies still emerging. Deploying today while inventing tomorrow is the only way to close the gap between ambition and achievement.—**Munir M. Eldesouki**

In addition to policy instruments facilitating transition, we need financial investments both in R&D and in the dissemination of emerging technologies. Public and private funding for research and development (R&D), demonstration projects, and early-stage commercialization is essential for advancing emerging technologies such as renewables and hydrogen, batteries, advanced nuclear systems, CCS, CCU, and DAC. Additionally, governments may provide concessional financing, loan guarantees, or direct public investment in low-carbon infrastructure. Climate finance provided by multilateral development banks, climate funds (e.g., the Green Climate Fund), and bilateral aid supports mitigation and adaptation efforts in developing countries.

The key is the institutionalization of Responsible Research and Innovation (RRI) in the financial sector. RRI is a governance framework that seeks to align research and innovation processes with societal needs, ethical standards, and sustainability goals. It emphasizes that scientific and technological development should not occur in isolation but rather through continuous interaction with society. RRI is commonly understood through key dimensions such as anticipation, inclusion, reflexivity, and responsiveness. Anticipation refers to the systematic assessment of potential impacts, risks, and uncertainties associated with innovation. Inclusion highlights the importance of engaging diverse stakeholders—including citizens, policymakers, industry, and civil society—in the research process. Reflexivity requires researchers and institutions to critically examine their own assumptions, values, and responsibilities. Responsiveness involves the capacity to adapt research directions in response to societal feedback and emerging knowledge.

In the financial sector, financial mechanisms and capital market innovations have become increasingly central to the decarbonization agenda. Green bonds, sustainability-linked bonds, and climate-aligned loans channel private capital into environmentally beneficial projects. Environmental, Social, and Governance (ESG) investment frameworks guide institutional investors toward low-carbon portfolios, while climate disclosure standards—such as those promoted by the Task Force on Climate-related Financial Disclosures (TCFD)—enhance transparency and risk assessment. Central banks and financial regulators are also integrating climate-related risks into macroprudential policy, stress testing, and supervisory frameworks.

However, contributions and responsibilities of investments and finance are still underestimated. An empirical analysis of investment-related Scope 3 emissions across the global financial system was conducted, addressing significant limitations in conventional carbon accounting approaches (Mejia & Kajikawa, 2026). Scope 3 emissions, particularly those associated with financial investments, represent a substantial share of total emissions but are often underestimated due to methodological constraints. Traditional accounting frameworks, such as those based on the Greenhouse Gas Protocol, typically attribute emissions according to direct ownership shares, thereby neglecting the complex and indirect relationships embedded in global investment networks. In particular, financial institutions—including asset managers and trust banks—emerge as significantly more carbon-intensive when network effects are considered. For example, large global financial actors exhibit dramatically higher investment-related emissions than previously estimated, reflecting their central role in allocating capital to carbon-intensive industries. Conversely, traditional industrial emitters, such as energy and manufacturing firms, show relatively lower attributed emissions under the network-based method, as their direct emissions are partially redistributed across the broader investment network (Mejia & Kajikawa, 2026). Therefore, in addition to harmonization of global policy in public sectors, responsible investment and finance in private sectors and institutions facilitating their responsible actions are crucial.

## 7 Orchestration of Business and Policy Ecosystems for System Transformation

> Colleagues, the journey to a zero-carbon society is not only a technological transition. It is a test of our collective will. It demands that we act now while investing in the breakthroughs of tomorrow. It demands courage, creativity, and collaboration.—**Munir M. Eldesouki**

To enable transformation to a zero-carbon society, we need an orchestration of the business and policy ecosystem. There are four fundamental roles of the ecosystem orchestrator (EO) (Gupta et al., 2024).

The first is leading value and stakeholder orchestration. Stakeholders generally agree with the necessity of system transformation but disagree with specific points. Thus, EO must design not only the societal and environmental values accompanying but also the values, merits, and necessities for each stakeholder. Based on value analysis, design, and orchestration at the systemic and stakeholder levels, EO must facilitate dialogues and coordination among them. It is not a rudimentary task to design relevant narratives to different stakeholders by considering different contexts ranging from global sustainability, business opportunities and feasibility, social acceptance of uncertain technologies, industrial competitiveness, geopolitics and security, and social welfare and justice.

The second is cultivating dynamic and interorganizational capabilities. Dynamic capability (DC) is for sensing, seizing, managing threats, and reconfiguring operational capabilities (Teece et al., 1997). But cultivating an individual firm's DC is not enough. EO must cultivate inter-organizational DCs by orchestrating network resources and governing network members. The realization of the implementation and commercialization of emerging technologies like CCS-CCU, DAC, and hydrogen energy depends on firms' ability to develop, deploy, and continually refine DCs across sensing, seizing, and reconfiguring dimensions. But it must accompany the transformation of supply chains through strategic integration, digitalization, collaborative partnerships, and circularity. And thus, the development of interorganizational DCs along value chains is critical. Firms that successfully cultivate dynamic capabilities and align them with supply chain innovation can accelerate the commercialization of emerging technologies, mitigate investment risks, and establish long-term competitive advantage in the global zero-carbon economy.

The third is interplaying innovation and business ecosystems. It stresses the difference between value creation in an innovation ecosystem and appropriation of value in a business ecosystem, and the importance of the design of the interplay of these different modes of ecosystems. Otherwise, we face the difficulties in finding offtakers of sustainable products and services after developing and accruing technologies. Advanced Market Commitments (AMCs), which are policy instruments designed to stimulate the development and production of products by guaranteeing a future market at a predetermined price, are one of the promising approaches. The central idea of AMCs is to reduce investment risk for private-sector producers in areas where demand is uncertain or where the financial incentives are insufficient for commercial viability. By providing a credible promise of future purchases, AMCs encourage firms to allocate resources to research, development, and manufacturing capacity that they might otherwise avoid due to high upfront costs and uncertain returns. Symbiosis Coalition by Google, Meta, Microsoft, Salesforce, McKinsey, Bain, and Rei Co-op is an example of AMCs, which commit to the future purchase of carbon removal credits.

The fourth is utilizing software ecosystems to integrate the heterogeneous needs of the stakeholders and to work as reference architectures for leveraging activities. We need various open-source software systems for navigating and enabling responsible actions, including monitoring and transparency systems along supply chains, transparent valuation and pricing systems along financial chains, and monitoring and assessment systems for global climate change.

> Let me share a recent example. During this year's pilgrimage season, welcoming almost 2 million pilgrims to the Kingdom, AI platforms developed by KACST provided real-time monitoring and decision support. These tools improved crowd safety while reducing the event's overall environmental footprint. It shows how digital innovation can deliver sustainability in practice, even in the most complex human settings.—**Munir M. Eldesouki**

We wish and believe that the STS *forum* has continuously served as an EO to facilitate the establishment of a network of networks among key stakeholders through scientific and transparent dialogues.

**Competing Interests** No competing interests.

## Session Panelists

**Wince-Smith, Deborah L. (Chair)**
President & CEO, Council on Competitiveness; President, Global Federation of Competitiveness Councils, U.S.A.
Deborah L. Wince-Smith is the President & CEO of the Council on Competitiveness, the Nation's premiere competitiveness leadership organization, representing U.S. CEOs, university presidents, and labor leaders. She leads national & international initiatives that assess competitiveness challenges, convene leaders who can envision and implement solutions, and organize action to enhance U.S. competitiveness. For most of her career, she has worked on a wide range of science, technology, innovation, & competitiveness issues advising top-level government policy-makers & business leaders. She serves on the boards of several public & private organizations. She is the President of the Global Federation of Competitiveness Councils, and a member of Japan's Science and Technology in Society forum Council.

**Kanda, Masato**
President and Chairperson of the Board of Directors, Asian Development Bank, Philippines
Masato Kanda is the President of the Asian Development Bank and Chair of its Board of Directors, having assumed office on 24 February 2025. He previously served as Special Advisor to Japan's Prime Minister and Minister of Finance. For nearly four decades at Japan's Ministry of Finance, Mr. Kanda held various roles in international affairs, budget management, financial services, and policy planning and coordination. As Vice-Minister of Finance for International Affairs, he played a key role at the G7 and G20, addressing issues such as multilateral development bank evolution, pandemic response, and debt sustainability. He also served as Chair of the OECD Corporate Governance Committee and as Alternate Executive Director for Japan at the World Bank.

**Eldesouki, Munir M.**
President, King Abdulaziz City for Science and Technology (KACST), Saudi Arabia
He is the President of KACST, and was the head of the establishment team of the Research Development and Innovation Authority. He was the Assistant Minister of Communications and Information Technology, the Deputy Minister of Planning and Development, and Senior Advisor on various Vision 2030 programs, leading and in establishing new government entities and transforming existing ones. He's also the Vice Chair of the Governing Board for Global Research Council. In recognition of his scientific excellence, he received several awards such as the King Abdulaziz Order of Merit Excellence Medal of the First Class, the King's Prize for Inventors, the Natural Science and Engineering Research Council of Canada scholarship, and was elected as a fellow of the National Academy of Inventors in the USA.

**Imauven, Claude**

Chairman of the Board of Directors, Orano, France

Claude Imauven is Chairman of the Board of Directors of Orano since 2020. He was, in the past, a board member of Orano Group. Previously, Claude Imauven held several positions for Saint Gobain company. He was Chief Operating Officer from 2016 to 2019. He was EVP in charge of the Construction Products Division from 2004 to 2015. Earlier, he was Chairman & CEO of Saint-Gobain PAM and managed the Pipe Division (2001-2004). He was Saint Gobain's General Delegate for Spain, Portugal & Morocco from 1996 to 2001. He was VP finance and VP of Industrial Policy of the Group's Flat Glass Division. Formerly, he worked for French Public administration where he held several positions of responsibility. He is a graduate of the École Polytechnique engineering school and also "Ingénieur au corps des Mines".

**Al-Khowaiter, Ahmad O.**

Executive Vice President, Technology & Innovation, Aramco, Saudi Arabia

Ahmad O. Al-Khowaiter was appointed Executive Vice President of Technology & Innovation on April 1, 2023. Prior to this role, he served as Senior Vice President and Chief Technology Officer for Aramco, a role he undertook in May 2015. He was previously Executive Director of Power Systems in 2014, and Chief Engineer from 2011 to 2014. Ahmad holds a B.S. degree in Chemical Engineering from the King Fahd University of Petroleum & Minerals (KFUPM), and a M.S. degree in Chemical Engineering from the University of California at Santa Barbara, as well as an M.S. degree as a Sloan Fellow from the Massachusetts Institute of Technology.

**Matsuo, Takehiko**

Vice-Minister for International Affairs, Ministry of Economy, Trade and Industry (METI), Japan

36 years of experience working for the Ministry of Economy, Trade and Industry (METI) as well as Cabinet Office and other agencies. Mr. Matsuo Takehiko has been serving as Vice Minister for International Affairs at the Ministry of Economy, Trade and Industry (METI) since July 2024. He leads all of METI's international policy operations and represents the ministry in cross-border economic dialogues and negotiations. Mr. Matsuo graduated from the University of Tokyo with a degree in law and from the Paul H. Nitze School of Advanced International Studies (SAIS) at Johns Hopkins University, with a master's in international public policy.

**Pathumnakul, Supachai**

Permanent Secretary, Ministry of Higher Education, Science, Research and Innovation, Thailand

Prof. Dr. Supachai Pathumnakul is the Permanent Secretary of the Ministry of Higher Education, Science, Research and Innovation (MHESI), Thailand. MHESI plays a vital role in driving Thailand's higher education, research, and innovation ecosystem, with a mission to strengthen human capital, promote science and technology, and support national strategies. Prof. Dr. Supachai earned his Bachelor's degree in Industrial Engineering from Khon Kaen University, and both Master's and Doctoral degrees in Industrial Engineering from Iowa State University, U.S.A. Throughout his career, Prof. Dr. Supachai has held several leadership roles in higher education and research management. Before assuming his current position, he served as Deputy Permanent Secretary of MHESI from 2020 to 2024. At Khon Kaen University, he was Vice President for Innovation and Enterprise, Vice President for Research and Technology Transfer, and Director of Khon Kaen University Science Park. He also served as President of the Research University Network of Thailand.

**Kato, Akihiko**

President & CEO, Atomic Energy Association (ATENA), Japan

Akihiko Kato has been President and CEO of Atomic Energy Association (ATENA), Japan, since June 2025. He previously had 40 year experience as an engineer and the senior management of Mitsubishi Heavy Industries (MHI), Ltd. He especially contributed to operators' efforts of restarting 12 PWR units in Japan with intensively supporting the deployment of various safety measures after the accident of Fukushima Daiichi Nuclear Power Plant. He also engaged in the design development of MHI's original next generation PWR, "SRZ-1200." He graduated from Kyushu University Graduate School with majoring in energy conversion engineering in 1984.

## References

Anderson, R., Bayer, P. E., & Edwards, D. (2020). Climate change and the need for agricultural adaptation. Current Opinion in Plant Biology, 56, 197–202.

Axon, C. J., & Darton, R. C. (2021). Sustainability and risk – A review of energy security. Sustainable Production and Consumption, 27, 1195–1204.

Dawood, F., Anda, M., & Shafiullah, G. M. (2020). Hydrogen production for energy: An overview. International Journal of Hydrogen Energy, 45(7), 3847–3869.

Ebi, K. L., Capon, A., Berry, P., Broderick, C., de Dear, R., Havenith, G., Honda, Y., Kovats, R. S., Ma, W., Malik, A., Morris, N. B., Nybo, L., Seneviratne, S. I., Vanos, J., & Jay, O. (2021). Hot weather and heat extremes: Health risks. The Lancet, 398(10301), 698–708.

Grape, S., Svärd, S. J., Hellesen, C., Jansson, P., & Lindell, M. Å. (2014). New perspectives on nuclear power—Generation IV nuclear energy systems to strengthen nuclear non-proliferation and support nuclear disarmament. Energy Policy, 73, 815–819.

Gupta, R., Mejia, C., Sadreghazi, S., Sano, Y., Muchangos, L. S. D., Sekiguchi, T., Nakamura, H., & Kajikawa, Y. (2024). Traders as ecosystem orchestrators for sustainable food supply chain transformation. Business Strategy and the Environment, 33(5), 4024–4038.

Huang, H. H., Kerstein, J., & Wang, C. (2018). The impact of climate risk on firm performance and financing choices: An international comparison. Journal of International Business Studies, 49(5), 633–656.

Kates, R. W., & Parris, T. M. (2003). Long-term trends and a sustainability transition. Proceedings of the National Academy of Sciences, 100(14), 8062–8067.

Klenert, D., Mattauch, L., Combet, E., Edenhofer, O., Hepburn, C., Rafaty, R., & Stern, N. (2018). Making carbon pricing work for citizens. Nature Climate Change, 8, 669–677.

Lau, H. C., Ramakrishna, S., Zhang, K., & Radhamani, A. V. (2021). The role of carbon capture and storage in the energy transition. Energy & Fuels, 35(9), 7364–7386.

Mathiesen, B. V., Lund, H., Connolly, D., Wenzel, H., Østergaard, P. A., Möller, B., Nielsen, S., Ridjan, I., Karnøe, P., Sperling, K., & Hvelplund, F. K. (2015). Smart energy systems for coherent 100% renewable energy and transport solutions. Applied Energy, 145, 139–154.

Mejia, C., & Kajikawa, Y. (2026). Estimating scope 3 emissions in the global investment network. Sustainability Science, 21(1), 39–57.

Shahbaz, M., Nasir, M. A., Hille, E., & Mahalik, M. K. (2020). UK's net-zero carbon emissions target: Investigating the potential role of economic growth, financial development, and R&D expenditures based on historical data (1870–2017). Technological Forecasting and Social Change, 161, 120255.

Teece, D. J., Pisano, G., & Shuen, A. (1997). Dynamic capabilities and strategic management. Strategic Management Journal, 18(7), 509–533.

Vakulchuk, R., Overland, I., & Scholten, D. (2020). Renewable energy and geopolitics: A review. Renewable and Sustainable Energy Reviews, 122, 109547.

Welton, S., & Harrison, C. (2025). Lessons in climate derisking: The United States' failed nuclear renaissance. University of Pennsylvania Law Review, 173(3), 705–788.

# Chapter 6
# Sustainability for Nature Positive Economy

**Yuya Kajikawa**

## 1 From Carbon Neutral to Nature Positive

> This is a very clear, alarming signal that we are totally out of balance with the natural system. As you all know, the natural system is the foundation of our humanity, prosperity, and society. —**Naoko Ishii.**
>
> The entire ocean is warming—not just the surface, but even the deepest portions of the ocean, even 6000 to 8000 meters deep. In addition, we are seeing more and more of the kinds of heat waves we experience on land occurring in the ocean. In 2023, one-third of the surface waters—the upper, say, 200 to 500 meters of the ocean—experienced marine heat waves. Those are times when the average surface temperature is more than 2 standard deviations above normal for more than 5 days. That's extraordinary—one-third of the ocean. —**Margaret Leinen.**

Reflecting the Great Acceleration in Anthropocene, including exponential increase of population, economy, production, and consumption (Komiyama & Kraines, 2008; Steffen et al., 2015a), economic burdens of our activities exceed the carrying capacity of the planet, so-called planetary boundaries (Steffen et al., 2015b; Richardson et al., 2023). The planetary boundaries framework is a scientific conceptualization of a safe operating space for humanity, which is defined in relation to critical biophysical processes that regulate the stability and resilience of the Earth system. It assumes that the Earth behaves as a complex, nonlinear system with tipping points, beyond which abrupt, potentially irreversible changes may occur. These thresholds are not always precisely known; therefore, the framework defines precautionary boundaries positioned at a safe distance from estimated tipping points. Crossing these boundaries increases the risk of destabilizing the Earth system, potentially leading to states less conducive to human well-being.

The updated planetary boundaries framework shows that six of the nine boundaries are transgressed, suggesting that Earth is now well outside of the safe operating

Y. Kajikawa (✉)
Institute for Future Initiatives, The University of Tokyo, Tokyo, Japan


H. Komiyama et al. (eds.), *AI, Innovation, and Beyond*,
https://doi.org/10.1007/978-981-95-7020-1_6

space for humanity (Richardson et al., 2023). It includes climate change ($CO_2$ concentration and radiative forcing), biosphere integrity (genetic and functional biodiversity), land system change, freshwater change, and novel chemical entities. Therefore, in addition to the combat with climate change and ambition to achieve a zero-carbon society and reduction of environmental damage by novel chemical entities like microplastics, Per- and Poly Fluoroalkyl Substances (PFAS), Organophosphorus Ester compounds (OPEs), nature positive relating to biosphere integrity, land system change, freshwater change is our common and essential challenge to safeguard the global commons. However, recovering stratospheric ozone levels and reduced atmospheric aerosol loading are strong signs that past environmental policy has been effective. Thus, there is potential for relevant nature-positive policies to recover and restore nature.

Nature positive is becoming a global agenda. For example, Kunming–Montreal Global Biodiversity Framework (GBF) emerged as a response to the accelerating global biodiversity crisis and the recognized shortcomings of prior international biodiversity governance frameworks. It was formally adopted at the 15th Conference of the Parties (COP15) to the Convention on Biological Diversity in 2022. GBF states that the integrity, connectivity, and resilience of all ecosystems are maintained, enhanced, or restored, substantially increasing the area of natural ecosystems by 2050. It also aims to halt human-induced extinction of known threatened species, reduce the extinction rate and risk of all species by 10-fold, and increase the abundance of native wild species to healthy, resilient levels by 2050.

The historical context of the GBF is rooted in the trajectory of international biodiversity policy since the adoption of the Convention on Biological Diversity (CBD) in 1992. The CBD established three overarching objectives: the conservation of biological diversity, the sustainable use of its components, and the fair and equitable sharing of benefits arising from genetic resources. Despite these commitments, biodiversity loss has continued at an unprecedented rate. Currently, there is a scientific consensus reflected in assessments such as those of the Intergovernmental Science-Policy Platform on Biodiversity and Ecosystem Services (IPBES), which indicated that approximately one million species were at risk of extinction, largely due to anthropogenic drivers including land-use change, overexploitation, pollution, invasive species, and climate change.

A critical precursor to the GBF was the Aichi Biodiversity Targets, adopted in 2010 under the CBD Strategic Plan for Biodiversity 2011–2020. These 20 targets aimed to address key drivers of biodiversity loss and improve conservation outcomes. However, global evaluations revealed that none of the Aichi Targets were fully achieved by 2020. This failure was attributed to several factors, including insufficient political commitment, lack of integration across sectors, inadequate financing, weak monitoring systems, and the absence of enforceable mechanisms. The perceived failure of the Aichi Targets created strong momentum for a more robust, measurable, and actionable post-2020 framework.

The emergence of voluntary and multi-stakeholder initiatives also reflects the growing role of non-state actors. For instance, the Science Based Targets Network (SBTN), launched in 2019, provides methodologies for companies to set science-based targets for nature, extending the logic of climate targets to biodiversity and ecosystems. Similarly, the Taskforce on Nature-related Financial Disclosures (TNFD), established in 2021, seeks to integrate nature-related risks and dependencies into financial disclosure. Global Environment Facility (GEF), established in 1991, serves as a key funding instrument for biodiversity conservation. More recently, new funding mechanisms have been proposed or established to meet the biodiversity finance gap. One mechanism is TFFF (Tropical Forest Forever Facility), launched at COP30, which is among the most significant recent innovations in global climate and biodiversity finance.

Compared with the policy instruments for climate change and a zero-carbon society, those for a nature-positive society are complex and at a nascent stage of development. There are sequential necessary steps for making nature more positive and realistic, including monitoring, assessment, and valuation of the ecosystem and its functions. Policy interventions to enable the monetization of nature-positive activities are also necessary, as illustrated below.

## 2 Monitoring of Nature and Assessment of Ecosystem Functions

The first step is to monitor the status of nature and to assess the function of ecosystem services. Monitoring of land-use change is a mature technique. Satellite imagery is a powerful tool for monitoring land-use and land-cover changes over time. The process typically involves acquiring multi-temporal images from sensors, which provide spectral information across visible, near-infrared, and shortwave-infrared bands. Preprocessing steps include geometric correction, radiometric calibration, and cloud masking to ensure consistency and comparability between images.

Land-use change is commonly assessed using classification methods that assign each pixel to a land cover category. Supervised classification relies on training data from known land cover types, while unsupervised classification groups pixels based on spectral similarity. By analyzing chronological change, we can quantify land-use transitions such as forest-to-agriculture conversion or urban expansion. This approach compares classified images from different years to generate change matrices and spatial maps of land cover dynamics. Accuracy assessment using ground truth or high-resolution reference data is essential for validating results and quantifying uncertainties.

Biodiversity monitoring is still under development while research is active. For example, forest biodiversity is inherently complex, encompassing species composition, ecosystem structure, and ecological processes across multiple spatial and temporal scales. Traditional biodiversity assessment methods, while precise, are often limited by cost, labor intensity, and spatial coverage, recent advances in Measurement, Reporting, and Verification (MRV) technologies such as environmental DNA (eDNA), satellite remote sensing, unmanned aerial vehicles (UAVs), acoustic monitoring, camera trapping, alongside traditional taxonomic surveys have significantly enhanced the capacity to monitor biodiversity (Lumbierres et al., 2025).

Based on monitoring and assessment using an integrative approach, we can investigate functions of ecosystems and their contributions to our society. We can also analyze the effects of exploiting nature and of using nature-positive interventions, as illustrated in the case of the ocean below.

> With respect to biodiversity, the heat waves certainly affect the organisms in the ocean, but so do the other impacts of $CO_2$ emissions and other greenhouse gas emissions: acidification, deoxygenation. All are reducing species in sensitive environments—not only the ones we hear about most, like coral reefs and mangroves, but also many of the ocean's productive surface areas. On top of that, overexploitation from fishing is driving diversity loss, both by direct removal of species like sharks, wild salmon, and tuna—all the things we hear a lot about—but also the removal of these apex predators in the ocean is causing major ecological disruptions down the food chain that are also affecting biodiversity. —**Margaret Leinen.**

Ecosystems have provided critical functions, including enhancing biological productivity, regulating hydrological cycles, and mitigating natural disasters such as floods and landslides. Quantifying these ecosystem functions is essential for sustainable management of natural capital and policy-making. The monitoring and assessment method should be scalable, accessible, and granular to support the further application of these assessment results, such as in biodiversity credits (Ford et al., 2024).

For example, water is a naturally renewable resource that continuously circulates and replenishes itself. While storage in natural and artificial reservoirs can augment the supply of water accessible to human societies, assessments of water resources should prioritize the dynamic flow of water rather than static stocks, which is modeled by hydrological simulation (Oki & Kanae, 2006; Liu et al., 2017). The climate system establishes a limit on the rate at which renewable freshwater resources can be replenished. However, global water withdrawals remain below this theoretical maximum, and over two billion people experience significant water stress due to the uneven temporal and spatial distribution (Oki & Kanae, 2006). Significant challenges persist in effectively integrating factors such as green water (soil moisture), water quality, environmental flow needs, globalization, and virtual water trade into assessments of water scarcity (Liu et al., 2017). In addition, the variability of water availability and consumption across seasons and years highlights the importance of considering the temporal dimension when evaluating water scarcity. Water footprint analysis and life-cycle assessment of water are available and have been reviewed (Boulay et al., 2018).

Nitrogen emissions from agricultural fertilization, livestock production, industrial activities, and fossil fuel combustion play a critical role in environmental degradation through eutrophication, acidification, and greenhouse gas emissions. Assessing the environmental impacts of nitrogen requires a combination of monitoring, biogeochemical cycle analysis, and hydrological simulation (Hurrell et al., 2013). Assessing pollutant loads (e.g., nitrogen and phosphorus) and applying water quality criteria are used to calculate the dilution volume required to meet environmental standards in the model. Field measurements provide direct evidence of nitrogen inputs, outputs, and transformations in soil, water, and the atmosphere. Soil nitrate and ammonium concentrations, nitrogen leaching into groundwater, and nitrous oxide ($N_2O$) fluxes are commonly quantified using chemical analyses, gas flux chambers, and isotopic tracing methods. Water quality monitoring, including nutrient concentrations in rivers and lakes, helps evaluate eutrophication risks and nutrient transport across landscapes.

## 3 Valuation of Natural Capital

> This has some really quite perverse consequences, because businesses put a price on nature, and that price today is zero. That's how decisions are made. —**Guido Schmidt-Traub.**
>
> Someone might say, 'Let's look at what crypto finance has told us,' because today, if you look at which assets are at their highest prices—literally today—gold is one, Bitcoin is another. I know that Pokémon cards are also doing very well at the moment. But why? From an economics point of view, none of them has intrinsic yield. —**Lee Howell.**

Natural capital encompasses the stock of natural resources and ecosystems that generate a wide array of ecosystem services essential to human well-being and economic activity. Assigning monetary value to natural capital has become increasingly important for integrating environmental considerations into policy, accounting, and investment decisions. In current business activities, nature is often viewed as infinite and free, leading to the belief that it can be wasted without payment or penalty. However, nature is in fact finite. The path to achieving nature positive lies in measuring and valuing this finite natural capital and integrating it into economic transactions and decision-making. Valuation involves stages such as qualitative, quantitative, and monetary valuation; while the latter is more difficult, it is also the most effective. A recent study of the Ecosystem Services Valuation Database registered information from 1300 studies, yielding over 9400 value estimates in monetary units for ecosystem services provided by 15 terrestrial and marine biomes (Brander et al., 2024). Monetary valuation is not intended to commodify nature but rather to make its contributions visible within predominantly economic decision-making frameworks. By assigning monetary values, policymakers can compare environmental benefits and costs with other economic activities, thereby facilitating more informed and sustainable choices.

Valuation consists of three approaches: cost-based, market-based, and income-based.

Cost-based methods estimate the value of ecosystem services based on the costs avoided or incurred in their absence. The avoided cost method calculates the costs that would have been incurred in the absence of the ecosystem service (e.g., flood damage avoided by wetlands). The replacement cost method estimates the cost of replacing ecosystem services with artificial alternatives (e.g., water treatment plants). Restoration cost methods assess the expenses required to restore degraded ecosystems. These approaches are practical and often used in policy contexts.

However, they may not accurately reflect societal preferences and can either overestimate or underestimate true values.

Market-based methods rely on observable market prices to estimate the value of ecosystem goods and services. These methods are most applicable to provisioning services such as timber, fish, and freshwater. The simplest approach is the market price method, which multiplies the quantity of goods by their market price. A more refined approach is the resource rent method, which calculates net benefits by subtracting production costs from revenues. While market-based methods are relatively straightforward and data-driven, they are limited in scope because many ecosystem services—such as carbon sequestration or biodiversity—are not traded in markets. Additionally, market prices may not reflect true social values due to distortions such as subsidies or externalities.

Income-based approaches have a variety of methods. For example, the travel cost method (TCM) estimates recreational value by analyzing how much individuals spend to visit natural sites. The hedonic pricing method examines how environmental attributes, such as air quality or proximity to green spaces, influence property prices. Stated-preference methods use surveys to elicit individuals' willingness to pay (WTP) or to accept (WTA) compensation for changes in ecosystem services. The contingent valuation method (CVM) directly asks respondents their WTP for specific environmental changes, while choice experiments present respondents with alternative scenarios and derive values from their choices.

Monetary valuation of natural capital provides a powerful tool for integrating environmental considerations into economic systems. However, it is not without limitations. Methodological challenges include uncertainty, data gaps, and difficulties in capturing complex ecological interactions. Ethical concerns also arise regarding the commodification of nature and the potential neglect of intrinsic values. To address these challenges, a pluralistic approach is recommended that combines monetary valuation with non-monetary indicators and participatory methods. Advances in data availability, remote sensing, and modeling are expected to improve the accuracy and applicability of valuation methods. More importantly, it is not always paid based on the valuation results, and valuation does not guarantee the monetary transactions.

## 4 Recognition and Registration of Natural Capital in Financial Statements

> Unless there's a global consensus and it enters into the System of National Accounts (SNA), it will not mean anything in the political process of individual countries. The SNA is the statistical national agreement on what every country's national statistical reporting will be. There have been attempts to advocate for that. Still, unless there is a standard agreement on what environmental accounting means, it will not be listened to by policymakers, particularly Ministers of Finance who are pushing for growth, growth, growth. —**Peter Gluckman.**

Today's national accounts capture only natural resources at market prices; the value of ecosystems and their services is largely missing. The UN System of Environmental-Economic Accounting (SEEA) provides a global standard to account for these, including monetary valuation using shadow prices. Over **90** countries compile accounts, but integration into fiscal and economic policy is rare. Where applied, results show how natural capital accounting can guide smarter investment and long-term resilience. This is seen with China's gross ecosystem product influencing budget signals, or Brazil's Plano Safra combining credits, insurance, and guarantees to steer agricultural finance. Once they account for nature, it will be clear to policymakers how many investment opportunities exist with positive returns.

> The biggest asset manager in the world, BlackRock, has said we need to be serious about natural capital, and we're starting to understand that half of global GDP depends on nature and the services it provides. But the reality is, if you look at some of the interesting products from someone like BlackRock, they own over 770,000 Bitcoins, worth about $96 billion in one of their investment funds right now. So serious money is being directed toward this asset that doesn't have an intrinsic yield. —**Lee Howell.**

In addition to the national account, implementing the financial accounts of companies is an ambitious challenge. There is scarce research and activity to realize recognition and registration of natural capital in financial statements, compared with accumulated research and publications on monitoring, assessment, and valuation of ecosystems, the functions of ecosystems, and ecosystem services. Ultimately, once natural capital is recorded on balance sheets as a new asset class, decisions that increase its value will enhance enterprise value (EV). In contrast, decisions that diminish it will lead to a decline in EV. Consequently, investing in natural capital and nature-based solutions that enhance its value will become an extremely economically rational decision.

> The tension is this is a lot of speculation—people wanting to get in and make a profit—and it's not in the spirit of stewardship that we're talking about when we discuss the planet. But if we're seeing disruption in finance, it is coming from the digital currency folks. They are not waiting for a global framework—it's really about arbitrage and finding where they can operate. I think we need to take a little bit of that playbook. —**Lee Howell.**

However, a natural asset company designed to hold and manage the rights to the productivity of natural assets, such as forests or wetlands, seeks an arbitrage and executes short-term trading, which apparently contradicts the aim of this instrument of nature on the balance sheet. Therefore, in addition to accounting rules, some regulatory policy is needed.

## 5 System Transformation and Transdisciplinary Science for Nature Positive Economy

> Public procurement, not only in the European setting but throughout the world, is a completely underused engine when it comes to innovation and making this shift. —**Mari Sundli Tveit.**

There are several ways to allow monetization and cash flow to natural capital. These policy options include payment for ecosystem services, public procurement, a nature tax, and biodiversity- and water-positive credits. A typical one is biodiversity net gain (BNG) and biodiversity credit in the UK. BNG policy emerged in response to long-standing concerns about biodiversity loss driven by land-use change and development. The historical background of BNG in the UK is overviewed by Stuart et al. (2026). BNG is a "net outcome" planning policy that aims for development projects to leave biodiversity in a better state than before they started. It also allows offsets by biodiversity credits. The relevance of offsets and the policy's efficiency, efficacy, and effectiveness will be studied, but this is an essential step.

System transformation toward a sustainable society involves a multilayer transformation (Geels, 2002). It is not limited to implementing alternative policies such as BNG and biodiversity credits, also includes development and deployment of nature-based solutions, investment in natural capital, business model transformation, business ecosystem transformation, reconsidering social infrastructure, lifestyle and behavioral change, etc. We have plenty of literature on those topics. But a fundamental question is whether this research contributes to accelerating system transformation.

> The amount of science that has been done in the last ten years in the name of the SDGs has not led to progress in the SDGs. Yes, technology has offered some routes forward, but science as a whole has failed. Why has it failed? —**Peter Gluckman.**

Since the 1990s, research on sustainability has experienced exponential growth in volume. But it is confined within disciplinary silos such as agriculture, forestry, fisheries, energy, biodiversity, water, health, urban development, and rural sociology structure. Since the 2010s, these silos have been integrated into environmental systems, the economy, and business systems through interdisciplinary research with a systemic perspective (Kajikawa et al., 2014). However, we must distinguish iinterdisciplinary research from transdisciplinary research (Tress et al., 2004). The latter accompanies common goal-setting, collaborative research, and actions by academic and non-academic bodies.

> If you read our reports—"Unleashing Science," 'Flipping the Science Model,' etc.—you'll see that science has stayed in silos, reduced to narrow disciplines, not engaging with stakeholders, whether communities or policymakers, and not taking a systems-based approach. If you look at the planetary boundaries framework, it's not a narrow, reductionist analysis—it's systems analysis. We in science largely don't do systems-based research, and certainly not systems-based research that engages with stakeholders. That's called transdisciplinarity. Transdisciplinary research takes a different approach and involves stakeholders in designing the question and the methodology. —**Peter Gluckman.**

Sustainability science has a normative approach, including visioning and goal setting, and a backcasting approach from the goal. It also has a pragmatic approach with causal and problem-solution chain analyses (Kajikawa, 2008). However, such normative, pragmatic, and action-oriented transdisciplinary approaches remain scarce. From a normative perspective, in addition to the economic incentives established in this chapter, vision and visioning are essential (Wiek & Iwaniec, 2014). Visions are not just imaginative descriptions of desirable futures, but structured tools that guide decision-making, stakeholder engagement, and transformative change. And visioning is a participatory process involving stakeholders, supporting collective learning, guiding policy and innovation, and helping societies move toward more sustainable futures.

Contrary to carbon neutral, which has global impacts, nature positive has primarily local impacts. Nature depends on local natural conditions, and ecosystem services and their utilization have historical, cultural, and social backgrounds in the region. And thus, the community-based participatory approach and landscape approach are meaningful, where inspired leadership, value proposition, long-term adaptive commitment, private sector engagement, systemic governance, and monitoring of metrics and process outcomes are needed (Sayer et al., 2015).

> We read the evidence, we hear the warnings, and we discuss the implications. Despite statements to the contrary by some figures—one in particular, I won't mention names—the fact remains, in my view, that we are fast approaching the point of no return unless concerted action is taken to address the decline by the world at large, by everybody. Systems must change, or we will transgress into the danger zone. The challenge, therefore, is to translate the conversations we have here today into collaboration and collective action. Let me state plainly that without fairness, our collective future will be bleak. —**Ahmad Tajuddin Ali.**

We wish the STS *forum* would serve as an arena for dialogue among diverse stakeholders and as a networked place for systemic design of sustainable transformation.

**Competing Interests** No competing interests.

## Session Panelists

| |
|---|
| **Ishii, Naoko (Chair)**<br>Special Presidential Envoy, Professor and Founding Director, Center for Global Commons, The University of Tokyo, Japan<br>Dr. Naoko Ishii is Professor and Director of the Center for Global Commons (CGC) at the University of Tokyo. Before joining UTokyo, she was the CEO and Chairperson of the Global Environment Facility (GEF) from 2012 to 2020. She is a vocal advocate for catalyzing systems change for sustainable development within planetary boundaries. She brings her expertise to global discourse by joining Independent High-Level Expert Group on Climate Finance, Global Commission on the Economics of Water (GCEW), High-Level Independent Panel on financing pandemic preparedness, among others. Her earlier role includes Deputy Vice Minister of Finance in Japan, senior positions at the World Bank and IMF. |
| **Gluckman, Peter**<br>President, International Science Council (ISC), France; Director, Koi Tu Centre for Informed Futures, New Zealand |
| **Schmidt-Traub, Guido**<br>Partner, Systemiq Ltd., France<br>Dr. Guido Schmidt-Traub is Partner of Systemiq Ltd, the global system change company. He was Managing Partner from 2021–2024. His work focuses on finance, adaptation, value chain and country transitions, working with investors, companies, governments,& civil society organizations. Guido serves on several boards. Previously he was Executive Director of the UN Sustainable Development Solutions Network, CEO of Paris-based CDC Climate Asset Management, Partner at South Pole Carbon Asset Management, Climate Change Advisor to Kofi Annan's Africa Progress Panel, head of the UNDP MDG Support Team, and Associate Director of the UN Millennium Project. He holds a PhD in Economics from Wageningen Univ., an M. Phil. in Economics from Oxford Univ., and a Masters in Physical Chem. from Free Univ. of Berlin. |

**Leinen, Margaret**

Director, Scripps Institution of Oceanography, University of California San Diego; Vice Chancellor, Marine Sciences, University of California, San Diego, U.S.A.

Margaret Leinen is the Director of Scripps Institution of Oceanography and Vice Chancellor for Marine Science of University of California at San Diego. Dr. Leinen is an ocean biogeochemist and paleoceanographer whose research includes study of ocean carbon cycling and the role of the oceans in climate. She is also Past President of the American Geophysical Union, and has also served as the President of The Oceanography Society and Chair of the AAAS Section on Atmospheric and Hydrospheric Science. She served as Assistant Director for Geosciences, U.S. National Science Foundation (NSF) from 2000–2007. She has been the Vice Chair of the International Geosphere Biosphere Programme, Chair of the US Global Change Research Program and Vice Chair of the U.S. Climate Change Science Program.

**Tveit, Mari Sundli**

Chief Executive Officer, Research Council of Norway (RCN), Norway; President, Science Europe, Belgium

Mari Sundli Tveit is Chief Executive of the Research Council of Norway. Sundli Tveit has previously been rector of the Norwegian University of Life Sciences and President of Universities Norway. Sundli Tveit studied nature management at the University of Tromsø, Bergen, and the Norwegian University of Life Sciences, from where she also holds a PhD and became a full professor of landscape architecture in 2013. Sundli Tveit has held a number of positions in the R&D sector, including on the board of the European University Association. She is president of Science Europe and member of the board of the Global Research Council. Her past position as policy director at the Confederation of Norwegian Enterprise has given her substantial insight into the R&I needs of business and industry.

**Howell, Lee**

Executive Director, Villars Institute, Switzerland

Professor Lee Howell is the Executive Director of the Villars Institute, a Swiss non-profit foundation advancing interdisciplinary research, intergenerational collaboration, and systems leadership. He is an executive and educator with a global, multi-faceted professional background which includes serving as Managing Board of the World Economic Forum and being responsible for the Annual Meeting in Davos (2009–2021). Lee is a Titular Professor at the Geneva School of Economics and Management at the University of Geneva. He also serves as the Vice Chair of the Board of Governors of the International Baccalaureate (IB) and is active in promoting scientific research as a board member of the EAT Foundation and as an advisory board member of the Frontiers Planet Prize.

**Ali, Ahmad Tajuddin**

Joint-Chairman, Malaysian Industry-Government Group for High Technology (MIGHT); Chairman, Universiti Tenaga Nasional (UNITEN) / The Energy University, Malaysia

Dr Ahmad Tajuddin Ali is currently the Joint-Chairman (Industry) of MIGHT, a government-industry think-tank for promoting high technology investments in Malaysia. He is also the Chairman of the Board of Linde Malaysia Holding Bhd, Plytec Bhd, ITMAX System Bhd and NuEnergy Holdings Bhd. He was previously the Executive Chairman of Tenaga Nasional Bhd and later Chairman of the Energy Commission Malaysia. In 1994, he was bestowed the Special Award by Japan External Trade Organization (JETRO) for his contribution towards enhancing exports of Malaysian products to Japan. In November 2018, he was conferred the award of "The Order of the Rising Sun, Gold Rays with Neck Ribbon" by the Government of Japan for his contribution towards promoting economic relations between Japan and Malaysia.

## References

Boulay, A. M., Bare, J., Benini, L., Berger, M., Lathuillière, M. J., Manzardo, A., Margni, M., Motoshita, M., Núñez, M., Pastor, A. V., Ridoutt, B., Oki, T., Worbe, S., & Pfister, S. (2018). The WULCA consensus characterization model for water scarcity footprints: Assessing impacts of water consumption based on available water remaining (AWARE). International Journal of Life Cycle Assessment, 23(2), 368–378.

Brander, L. M., de Groot, R., Schägner, J. P., Guisado-Goñi, V., van't Hoff, V., Solomonides, S., McVittie, A., Eppink, F., Sposato, M., Do, L., Ghermandi, A., Sinclair, M., & Thomas, R. (2024). Economic values for ecosystem services: A global synthesis and way forward. Ecosystem Services, 66, 101606.

Ford, H. V., Schrodt, F., Zieritz, A., Exton, D. A., van der Heijden, G., Teague, J., Coles, T., & Field, R. (2024). A technological biodiversity monitoring toolkit for biocredits. Journal of Applied Ecology, 61(9), 2007–2019.

Geels, F. W. (2002). Technological transitions as evolutionary reconfiguration processes: A multi-level perspective and a case-study. Research Policy, 31, 1257–1274.

Hurrell, J. W., Holland, M. M., Gent, P. R., Ghan, S., Kay, J. E., Kushner, P. J., Lamarque, J.-F., Large, W. G., Lawrence, D., Lindsay, K., Lipscomb, W. H., Long, M. C., Mahowald, N., Marsh, D. R., Neale, R. B., Rasch, P., Vavrus, S., Vertenstein, M., Bader, D., … Marshall, S. (2013). The Community Earth System Model: A framework for collaborative research. Bulletin of the American Meteorological Society, 94(9), 1339–1360.

Kajikawa, Y. (2008). Research the core and framework of sustainability science. Sustainability Science, 3, 215–239.

Kajikawa, Y., Tacoa, F., & Yamaguchi, K. (2014). Sustainability science: The changing landscape of sustainability research. Sustainability Science, 9(4), 431–438.

Komiyama, H., & Kraines, S. (2008). Vision 2050—Roadmap for a Sustainable Earth. Springer.

Liu, J. G., Yang, H., Gosling, S. N., Kummu, M., Flörke, M., Pfister, S., Hanasaki, N., Wada, Y., Zhang, X. X., Zheng, C. M., Alcamo, J., & Oki, T. (2017). Water scarcity assessments in the past, present, and future. Earth's Future, 5(6), 545–559.

Lumbierres, M., Milanovic, M., Beja, P., Bonn, A., Breeze, T. D., Brotons, L., Fernández, N., Junker, J., Liquete, C., Solheim, A. L., Ordóñez, A. M., Moreira, F., Santana, J., Shinneman, S., Smets, B., Pereira, H. M., Valdez, J. W., van Grunsven, R. H. A., & Kissling, D. (2025). Towards implementing workflows for essential biodiversity variables at a European scale. Global Ecology and Conservation, 62, e03699.

Oki, T., & Kanae, S. (2006). Global hydrological cycles and world water resources. Science, 313(5790), 1068–1072.

Richardson, K., Steffen, W., Lucht, W., Bendtsen, J., Cornell, S. E., Donges, J. F., Drüke, M., Fetzer, I., Bala, G., von Bloh, W., Feulner, G., Fiedler, S., Gerten, D., Gleeson, T., Hofmann, M., Huiskamp, W., Kummu, M., Mohan, C., Nogués-Bravo, D., … Rockström, J. (2023). Earth beyond six of nine planetary boundaries. Science Advances, 9(37), eadh2458.

Sayer, J., Margules, C., Boedhihartono, A. K., Dale, A., Sunderland, T., Supriatna, J., & Saryanthi, R. (2015). Landscape approaches: What are the pre-conditions for success? Sustainability Science, 10, 345–355.

Steffen, W., Broadgate, W., Deutsch, L., Gaffney, O., & Ludwig, C. (2015a). The trajectory of the Anthropocene: The Great Acceleration. Anthropocene Review, 2(1), 81–98.

Steffen, W., Richardson, K., Rockström, J., Cornell, S. E., Fetzer, I., Bennett, E. M., Biggs, R., Carpenter, S. R., de Vries, W., de Wit, C. A., Folke, C., Gerten, D., Heinke, J., Mace, G. M., Persson, L. M., Ramanathan, V., Reyers, B., & Sörlin, S. (2015b). Planetary boundaries: Guiding human development on a changing planet. Science, 347(6223), 1259855.

Stuart, A., Bond, A., Franco, A. M. A., Gerrard, C., Baker, J., ten Kate, K., Butterworth, T., Bull, J., & Treweek, J. (2026). How England got to mandatory biodiversity net gain: A timeline. Ambio, 55(4), 714–742.
Tress, G., Tress, B., & Fry, G. (2004). Clarifying integrative research concepts in landscape ecology. Landscape Ecology, 20, 479–493.
Wiek, A., & Iwaniec, D. (2014). Quality criteria for visions and visioning in sustainability science. Sustainability Science, 9(4), 497–512.

# Chapter 7
# Brain Health and Brain Augmentation

**Yousif Elsamani**

## 1 Opening Vision: The Renaissance of the Brain

> We are at a crossroad. The convergence of neuroscience and AI offers the possibility of radically improving lives … but it also brings profound responsibility. The brain is not just another organ—it is where identity, memory, and emotion reside. —**Harriet Wallberg**

Humanity stands at the threshold of what might rightly be called a renaissance of the brain—a period when neuroscience, technology, and ethics converge to redefine how we understand ourselves. For most of the twentieth century, "brain health" was a clinical term confined to neurology and psychiatry. Today it is expanding into a societal construct, linking cognitive vitality with education, productivity, sustainability, and global equity (Elkind et al., 2025). As Wang et al. emphasize, brain health must be viewed holistically—encompassing physical, cognitive, social, and emotional dimensions that collectively enable individuals to realize their potential (Wang et al., 2020). The World Health Organization has accordingly declared brain health a global policy priority, urging nations to align it with the Sustainable Development Goals (Kolappa et al., 2022).

> The human brain … is no longer only the subject of study in terms of disease and dysfunction, but increasingly it is the platform for enhancement, extension, and even transformation. —**Harriet Wallberg**

This transformation is not merely technological; it is paradigmatic. Eyre et al. describe it as the rise of Brain Capital—the idea that cognitive and emotional resources form the foundation of national prosperity and resilience (Eyre et al., 2024). Framing brain function as capital shifts policy discourse from treatment to investment, where nurturing creativity, empathy, and adaptive intelligence becomes

Y. Elsamani (✉)
Institute for Future Initiatives, The University of Tokyo, Tokyo, Japan


H. Komiyama et al. (eds.), *AI, Innovation, and Beyond*,
https://doi.org/10.1007/978-981-95-7020-1_7

as vital as infrastructure or energy security. Within this worldview, the brain is simultaneously a biological organ and a socio-economic asset.

Meanwhile, the scientific frontiers are providing unprecedented empirical depth. Bethlehem et al.'s *Nature* study established brain charts for the human lifespan, mapping structural variability across more than 123,000 MRI scans (Bethlehem et al., 2022). These "neural growth charts" are the cognitive equivalent of population health baselines, enabling early identification of atypical trajectories and anchoring what "normal" development means in a data-rich era. Such advances exemplify the quantitative turn in neuroscience—where longitudinal datasets and AI-driven analytics are merging to produce a high-resolution understanding of human cognition.

This intellectual expansion is mirrored by an exponential rise in scholarly output. A search of the Web of Science database using the core terms *"brain health," "cognitive resilience,"* and *"brain augmentation"* shows a dramatic growth in publications since 1990 (Fig. 7.1). The trend reflects not only advances in neuroimaging and AI but also the increasing convergence of medicine, engineering, and social science around the shared objective of enhancing cognitive function. The steep climb after 2015 coincides with the surge of AI-based neurotechnologies and the framing of brain health as a global development priority (Wang et al., 2020; Kolappa et al., 2022; Eyre et al., 2024).

The figure shows the number of academic publications containing brain health–related terms in their titles or abstracts. Publication counts have risen sharply since the early 2000s, with an inflection point around 2015. By 2025, annual publications exceeded 1800 records—more than a tenfold increase over two decades—illustrating the emergence of brain health and augmentation as a distinct, multidisciplinary research frontier. Data source: Web of Science Core Collection (accessed October

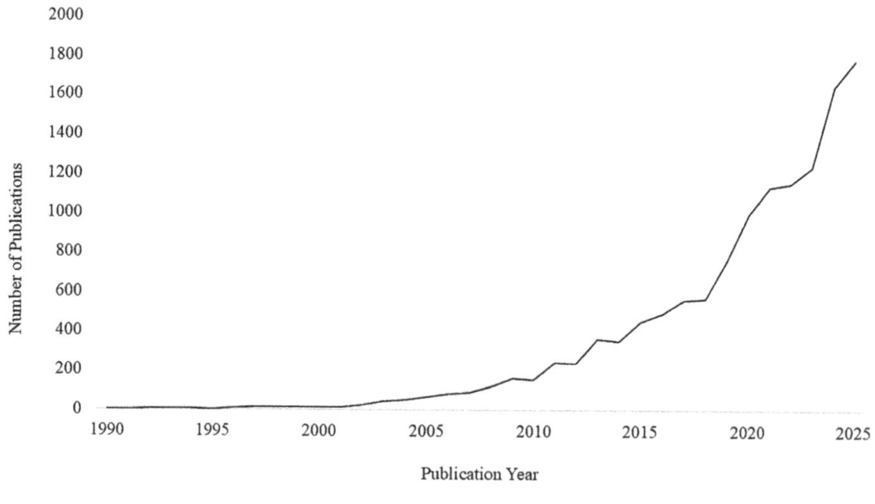


**Fig. 7.1** Temporal evolution of publications on brain health and brain augmentation (Web of Science, 1990–2025)

2025). Search query: TS = ("brain health" OR "cognitive resilience" OR "cognitive reserve*" OR "brain augmentation" OR "neuroaugmentation" OR "brain enhancement" OR "cognitive enhancement").

> The brain is the core of who we are … our task is to ensure that today's rapid advances serve humanity wisely and with responsibility. —**Harriet Wallberg**

In this sense, the renaissance of brain science is inseparable from the ethics of innovation. The challenge is to ensure that every algorithm trained to decode mental states or optimize learning outcomes ultimately strengthens, rather than erodes, the integrity of human thought. This book chapter begins by charting this intellectual landscape: how brain health became a unifying concept across medicine, economics, and sustainability; how the idea of Brain Capital redefined value in the cognitive age; and how the mapping of the human brain across the lifespan now anchors a new social contract between science and society.

## 2 From Restoration to Enhancement: The Expanding Frontier

> We've moved from observing the brain to shaping how it works—restoring movement, speech, sensation, and sometimes mood stability. —**Caroline Montojo**

The evolution of brain science over the past two decades has blurred the line between therapy and enhancement. What began as a quest to treat neurological disorders is rapidly transforming into an effort to augment healthy cognition and extend human capacity. Artificial intelligence, neuroengineering, and bioelectronic medicine have converged to form a continuum that links clinical rehabilitation to performance optimization.

AI-based systems now underpin most advanced neurotechnologies. Segato et al. reviewed how artificial intelligence is increasingly applied to diagnosis, prognosis, and personalized treatment of brain diseases, identifying deep learning as the core enabler of adaptive neuromodulation (Segato et al., 2020). Similarly, Zang et al. highlighted the rise of digital therapeutics—software-based interventions that monitor and modulate cognitive function outside traditional clinical settings (Zang et al., 2025). Together, these developments illustrate a paradigm shift: cognition itself has become programmable, measurable, and potentially upgradeable.

> Japan's Moonshot goal—liberating humans from the limits of body, brain, space, and time—demands neurotechnology that connects mind and world. —**Ryota Kanai**

Kanai's comment captures the aspirational horizon of this movement. From brain–computer interfaces (BCIs) to AI-powered prosthetics, restoration technologies are evolving toward direct neural co-processing, where human intention and machine learning operate as a unified loop. Liu et al.'s meta-analysis of BCI-assisted stroke rehabilitation demonstrates both the therapeutic efficacy and the expanding technical feasibility of decoding motor intentions from brain signals (Liu et al., 2025).

Each incremental improvement in decoding accuracy and feedback latency brings BCIs closer to everyday, non-clinical use—gaming, learning, and creative design among them.

Beyond individual trials, the field itself is maturing. Miao et al.'s *Lancet Digital Health* mapping of clinical trials in digital therapeutics shows an accelerating regulatory pipeline with standardized endpoints and risk frameworks (Miao et al., 2024). This formalization signals that enhancement is no longer speculative; it is entering the evidence-based domain traditionally reserved for medicine.

> As we push AI toward general intelligence, architectures may converge on consciousness—perhaps before we fully understand it. —**Ryota Kanai**

Such convergence raises a pivotal question: when cognitive restoration becomes indistinguishable from cognitive expansion, what exactly are we enhancing? The move from repairing neural pathways to redesigning them implies a new ethic of design—one that treats consciousness itself as a potential engineering problem. Yet the very success of these technologies also magnifies disparities in access, privacy, and agency. The next section turns to these ethical thresholds—where the science of augmentation meets the moral architecture of thought.

## 3 The Ethics of Thought: Privacy, Consent, and the Therapy–Enhancement Boundary

> Brain data is uniquely intimate—it reveals not just what we do, but how we think or feel. —**Caroline Montojo**

As brain–computer interfaces and AI-mediated neurotechnologies mature, the boundary between medical care and cognitive manipulation grows increasingly porous. The promise of neural restoration—re-enabling movement or speech—inevitably leads to a deeper question: who controls the mind once its signals can be decoded, transmitted, or optimized?

The traditional medical-ethics quadrants—autonomy, beneficence, non-maleficence, and justice—still provide the baseline compass for clinical research. Yet, as Baumann et al. argue, even these four principles falter when procedures extend beyond therapy into elective enhancement (Baumann et al., 2013). The patient–clinician dyad that once defined informed consent now meets a new triad: patient, algorithm, and data intermediary.

Society's fascination with neuro-enhancement has already surfaced in athletics, gaming, and military training, where neurostimulation and cognitive enhancers blur the line between skill and circuitry. Reardon's early report on "brain doping" in sports warned that performance optimization could outpace regulatory frameworks (Reardon, 2016). What once seemed fringe is now commercially available in wearable EEG headbands and transcranial stimulation devices marketed for focus, creativity, and emotional control.

This expanding accessibility underscores a broader challenge: technologies developed for medical recovery are increasingly crossing into recreational or competitive domains, raising complex ethical and social questions about fairness, consent, and human agency.

The Brain Capital framework proposed by Smith et al. repositions these dilemmas within a socio-economic lens: if cognitive potential is a form of capital, then the right to augment becomes a question of equitable access and national competitiveness (Smith et al., 2021). This logic risks converting mental capability into a marketplace commodity—valuable, tradable, and unequally distributed. Ibáñez and Eyre extend the argument further, linking ecological sustainability to cognitive well-being and calling for policies that balance neurotechnological progress with environmental and social integrity (Ibanez & Eyre, 2023). Such perspectives recast the ethics of augmentation as an issue of collective stewardship, not just individual choice.

> If my thoughts can be read, can we have a switch that protects intellectual property—because thoughts are free? —**One of the Audience at the Forum**

The plea from one participant at the STS session epitomized a growing public anxiety: the right to cognitive privacy. Emerging experiments in "neural decoding" underscore both the excitement and the peril of transparency. Montojo reminded the audience that even the most advanced BrainGate studies embed safety measures such as explicit verbal "code phrases" to initiate or halt data capture—a primitive but symbolic act of mental consent.

The ethical horizon is clear yet unfinished. As augmentation technologies scale, mental integrity—the capacity to think freely and remain the author of one's own cognition—must be recognized as a new human right.

The ethical horizon that emerged in the previous section also defines a conceptual crossroads. To move forward, neuroscience and society must reconcile three forces now shaping our collective future—the pursuit of *brain health*, the investment in *brain capital*, and the expansion of *brain augmentation*. Before turning to how these ideas are quantified, it is useful to visualize how they overlap conceptually and how their boundaries blur as innovation advances (Fig. 7.2).

The diagram illustrates the interconnections between three paradigms shaping the neuroscience and technology policy landscape. The horizontal axis represents the *therapeutic–enhancement continuum*, while the vertical axis reflects the shift from *individual* to *societal outcomes*. Brain Health emphasizes clinical restoration and prevention; Brain Capital reframes cognitive and emotional capacities as drivers of social and economic value; and Brain Augmentation extends human capability through neurotechnology. Their intersections highlight the need for responsible innovation that bridges medicine, economics, and ethics to foster inclusive and sustainable brain futures.

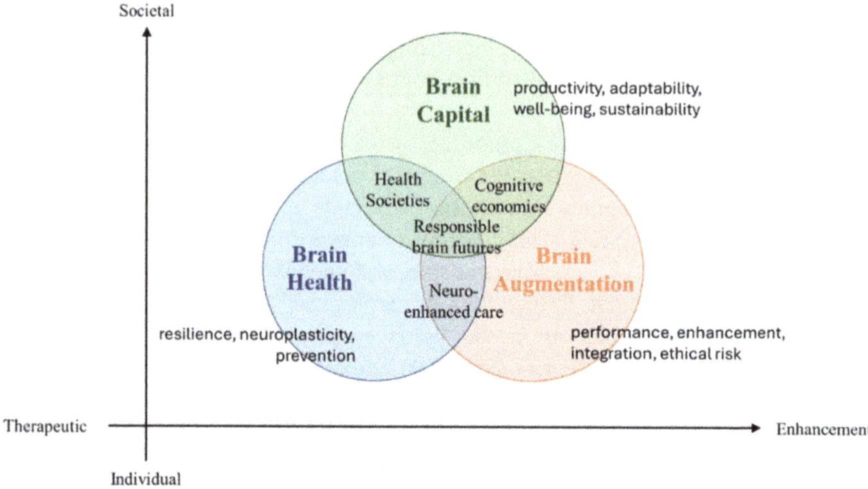


**Fig. 7.2** Conceptual integration of brain health, brain capital, and brain augmentation

## 4 Measuring the Mind: Quantifying Brain Health

> The Brain Health Quotient makes brain health readable—like IQ—so anyone can grasp their status and change over time. —**Yoshinori Yamakawa**

Modern neuroscience increasingly treats the brain not only as a source of thought but as a system that can be monitored, benchmarked, and optimized. The idea of a *Brain Health Quotient (BHQ)* reflects a broader scientific trend: translating complex neural and cognitive processes into quantifiable metrics accessible to individuals, clinicians, and policymakers alike. This quantitative turn offers a bridge between the lofty goals of brain augmentation and the practical realities of public-health management.

Brain-imaging and neurophysiological tools now capture subtler correlates of well-being than ever before. Functional near-infrared spectroscopy (fNIRS) studies have shown that even ordinary behaviors—such as moderate exercise—modulate cortical activation patterns linked to memory and executive function (Shen et al., 2024). Similarly, systematic reviews of physical activity across the lifespan confirm enduring relationships between aerobic fitness, neurogenesis, and cognitive performance (Voss et al., 2011). These results transform "lifestyle" into a measurable determinant of neural resilience, placing personal behavior within the scope of national brain-health strategy.

> We used to think brain aging began around sixty, but BHQ shows it starts declining near twenty—and, encouragingly, it can recover. —**Yoshinori Yamakawa**

Quantification is also accelerating through big-data synthesis. Bibliometric mapping of brain-atrophy research reveals rapidly converging interests across radiology, gerontology, and machine learning (Wang et al., 2024). Such meta-analyses expose the intellectual architecture of the field—how imaging biomarkers, cognitive-reserve theory, and AI diagnostics are coalescing into a unified language of measurement.

At the conceptual level, Harrison et al.'s review of *cognitive reserve* underscores why people with similar pathology exhibit divergent outcomes (Harrison et al., 2015): the mind's adaptability is as much about *capacity* as about *damage*. Embedding these constructs in quantitative indices like BHQ enables longitudinal tracking of resilience and recovery, transforming abstract neuroscience into actionable public indicators.

The policy implications are profound. Once brain health becomes measurable, it becomes governable—a target for investment, intervention, and accountability. BHQ initiatives in Japan, for example, illustrate how quantification can empower both individuals and institutions to pursue well-being as a shared societal goal rather than a private aspiration. The next section extends this discussion from measurement to meaning—how the design of brain technologies can promote inclusion, accessibility, and shared autonomy across diverse populations.

## 5 Designing for Inclusion: Accessibility and Shared Autonomy

> For people with visual impairments, augmentation starts as sensory substitution—converting vision into touch or sound. —**Chieko Asakawa**

At its most humane, brain augmentation is not about outpacing human limits but about restoring participation. The fusion of neuroscience and artificial intelligence has opened new avenues for inclusion, where technology acts as an equalizer rather than a divider. Dr. Asakawa's decades of work exemplify this transition—from translating visual data into tactile cues to creating AI-guided navigation for the visually impaired. Such systems demonstrate how the principles of brain augmentation—perception enhancement, autonomy, adaptive learning—can be deployed to enhance human dignity.

AI research has been pivotal in building these inclusive tools. Segato et al.'s systematic review shows that artificial-intelligence models now underpin most major innovations in the diagnosis and treatment of brain diseases (Segato et al., 2020). The same learning architectures that decode pathology are being repurposed for assistive cognition, enabling technologies that "think with" rather than "for" the user.

Zang et al. further highlight how digital therapeutics (DTx) are extending cognitive care into daily life—using adaptive feedback loops to improve memory, mood, and executive functioning outside the clinic (Zang et al., 2025). Together, these developments illustrate how the frontier of brain augmentation is moving from specialized laboratories to everyday environments, often mediated by smartphones, wearables, or cloud-based neural models.

> The AI Suitcase is moving from autonomy to shared autonomy—combining the robot's navigation with the user's moment-to-moment decisions. —**Chieko Asakawa**

Shared autonomy redefines the relationship between humans and machines. Instead of replacing human agency, intelligent systems now negotiate it—allocating control dynamically based on situational awareness, confidence levels, or emotional context. In Asakawa's work, for example, the AI Suitcase can decide when to lead and when to defer, creating an interactional rhythm that mirrors trust between companions. Such design principles echo broader findings in social neuroscience: Duffner et al.'s meta-analysis demonstrates that social health factors—community engagement, emotional bonds, and perceived belonging—are strongly associated with structural brain integrity and cognitive performance (Duffner et al., 2023). If relationships shape neural architecture, then inclusive technology must also nurture relationships—not just deliver functions.

> Privacy isn't abstract—if we ban the cameras these systems need, we take away independence from those who benefit most. —**Chieko Asakawa**

Designing for inclusion thus requires balancing ethical transparency with functional necessity. Many assistive technologies depend on sensors and cameras that inevitably capture bystanders, raising concerns over data privacy and surveillance.

Yet, as Asakawa reminds us, the cost of over-cautious regulation may be exclusion itself. Responsible innovation must therefore articulate the social value of accessibility and build frameworks where data ethics reinforce, rather than hinder, human empowerment.

The story of inclusive neurotechnology is not a niche narrative—it is a microcosm of what equitable augmentation could look like at scale. It reminds us that innovation must not only elevate capability but also expand compassion. The next section explores how these principles extend to the governance of neurotechnologies, where national policies, economic priorities, and ethical oversight converge.

## 6 Global Governance and Future Responsibility

As neurotechnologies migrate from laboratories to markets, their governance becomes both a scientific and moral imperative.

> Neurotechnology records and modulates the very activity that underpins freedom of thought. —**Didier Samuel**

If the twenty-first century belongs to any scientific frontier, it is the frontier of the mind. Yet the acceleration of neurotechnology—its medical, cognitive, and commercial uses—demands a governance framework as sophisticated as the science itself. Professor Didier Samuel's words capture a central paradox: the same technologies that promise healing and knowledge also grant access to the neural substrate of autonomy itself. Regulation, therefore, is not merely an administrative function; it is an act of safeguarding mental sovereignty.

The governance challenge begins with defining brain health as a societal and economic asset. Eyre et al. argue that "Brain Capital" is not a metaphor but a quantifiable form of value underpinning innovation, productivity, and well-being (Eyre et al., 2024). This notion places brain health at the heart of sustainable development strategies—connecting cognitive resilience to global issues such as education, mental health, and climate adaptation. In this light, fostering cognitive well-being is as vital to national competitiveness as energy infrastructure or digital connectivity.

Nail-Beatty et al. extend this argument to economic transitions, emphasizing that brain health is essential for stable adaptation during automation, demographic shifts, and climate disruptions (Nail-Beatty et al., 2024). They frame mental agility and creativity as strategic resources that help societies adjust to technological turbulence. This perspective aligns strongly with the Science and Technology in Society (STS) *forum*'s ethos—seeing technological innovation not as destiny, but as design: a collective choice that requires reflection, inclusion, and ethical foresight.

> AI is first and foremost a great source of progress—helping develop knowledge, diagnostics, and treatment—but it must be paired with rigorous oversight. —**Didier Samuel**

Global collaboration is central to effective oversight. Institutions such as the French National Institute of Health and Medical Research (INSERM), the U.S. National Institutes of Health (NIH), and the Organisation for Economic Co-operation and Development (OECD) have each contributed to emerging neuroethics frameworks, yet fragmented national approaches risk creating uneven ethical standards. As neural data flows across borders through cloud infrastructures and transnational research consortia, governance must evolve beyond traditional bioethics to embrace neurorights—protecting mental privacy, identity, and integrity at the policy level. This shift mirrors the recognition of digital rights two decades ago; today, it extends that protection into the cognitive domain. As Ayo Wahlberg observes, the imperative is to ensure that rapid advances serve humanity with wisdom and foresight.

Recent work on *Green Brain Capital* offers a vision for integrating ecological and cognitive well-being into policy design (Abdelraheem et al., 2025). Abdelraheem et al. propose metrics that align environmental resilience with mental health, linking biodiversity and climate stability to brain-function outcomes. Such interdisciplinary frameworks redefine sustainability: protecting ecosystems and protecting minds become mutually reinforcing goals.The result is a model of governance that treats human cognition as both beneficiary and steward of the planet's health.

As nations build their digital, biomedical, and cognitive infrastructures, the ethics of brain technology cannot remain a postscript. They must become embedded in economic policy, health systems, and education—ensuring that neurotechnology supports inclusion, autonomy, and planetary stewardship rather than amplifying inequity. The final section of this chapter turns to the intimate dimension of this transformation—the rhythms of the living brain itself, where light and shadow coexist within every act of consciousness.

In recent years, governments have begun taking the first steps toward protecting mental privacy and cognitive freedom in an era shaped by neurotechnology. Chile led the way in 2021, becoming the first country to enshrine neurorights in its constitution—recognizing the brain's data as an extension of personal identity and autonomy. Spain soon followed, adopting a national Charter of Digital Rights that includes explicit provisions for neurodata governance and human agency in human–machine interactions. At the international level, the OECD's Recommendation on Responsible Innovation in Neurotechnology (2019) calls on governments to anticipate ethical and legal challenges through proactive regulation, public engagement, and inclusive dialogue. Together, these initiatives signal a profound shift: from voluntary ethics—the traditional reliance on professional codes and self-restraint—toward rights-based governance, where the protection of thought and emotion becomes a matter of public policy.

## 7 Epilogue—Lights and Shadows

> The brain is the core of who we are. Our task is to ensure that today's rapid advances serve humanity wisely and with responsibility. —**Harriet Wallberg**

The story of brain health and brain augmentation is not a tale of triumph alone. It is a story of light and shadow—a narrative of discovery balanced by humility. The same tools that illuminate neural circuits also expose the fragility of consciousness; the same systems that expand cognition can, if misused, erode the very autonomy they promise to enhance.

Nowhere is this tension clearer than in the most elemental of all brain functions: sleep. As Hauglund et al. demonstrate, sleep activates the glymphatic system—a nocturnal "cleansing" mechanism that flushes metabolic waste and toxins from the brain (Hauglund et al., 2020). Without this quiet biological maintenance, neural efficiency and mental clarity degrade, regardless of how advanced our augmentation technologies become. Tai et al. extend this insight, linking sleep duration to executive function and measurable differences in brain structure (Tai et al., 2022). The finding is both physiological and philosophical: even the most advanced minds remain dependent on the simplest rhythms of rest and renewal.

In this sense, the pursuit of enhancement must coexist with reverence for limitation. While BCIs, neurostimulation, and AI-driven cognition redefine potential, they do not replace the organic cycles that sustain it. As Erickson et al. showed more than a decade ago, physical activity and lifestyle—mundane acts of care—still offer the most consistent protection against cognitive decline (Erickson et al., 2012). These empirical constants remind us that progress should not displace wisdom; it should deepen it.

> Technology should move in the direction of helping people live independently and with dignity. —**Chieko Asakawa**

Perhaps the greatest lesson from the STS *forum* is that the brain's future is inseparable from the human condition. Innovation, if it is to remain humane, must preserve not only function but meaning: the right to wonder, to rest, to forget, to dream. Between the neural light of progress and the shadow of ethical risk lies the path of responsibility—one illuminated not by technology itself, but by the values guiding its use.

As this chapter closes, the collective message of the session echoes: the next frontier of science is not to *build better brains*, but to deepen our understanding of what it means to be human in an age when our thoughts can be measured, modeled, and, perhaps one day, shared. In shaping the future of the brain, we are also shaping the moral architecture of the human condition.

**Competing Interests** No competing interests.

## Session Panelists

| |
|---|
| **Prof. Harriet Wallberg** was Director General at the Ministry of Health, Swedish Government 2016–2020. She was President of Karolinska Institutet (KI) between 2004–2012 and University Chancellor of Sweden between 2014–2016. She is a member of KI's Nobel Assembly selecting the recipient of the Nobel Prize in Physiology or Medicine. She earned her medical and doctoral degrees at KI. Harriet Wallberg has served as Secretary General of the Swedish Research Council in Medicine. She is member of the board of STS *forum*. She has received Honorary Doctorates from Univ. of Minnesota, USA 2010, Trinity College, Ireland 2011 and Seoul National Univ., Korea 2012. She has been honored with H. M. the King's Medal for distinguished achievements in medical education and research in 2008. |
| **Caroline Montojo, PhD**, is the President and CEO of the Dana Foundation, a private philanthropic organization in New York City dedicated to advancing neuroscience and society by supporting cross-disciplinary intersections such as neuroscience with ethics and law. Dr. Montojo is on the Science Philanthropy Alliance Advisory Board. She was previously the Director of Life Sciences at The Kavli Foundation. Dr. Montojo completed postdoctoral research in the Semel Institute for Neuroscience and Human Behavior at the University of California, Los Angeles. Dr. Montojo received her bachelor's degree from the University of California, Berkeley, and completed her Ph.D. at Johns Hopkins University. |
| **Kanai Ryota.** After graduating from Kyoto University's Faculty of Science (2000), he earned his PhD Cum Laude from Utrecht University (2005) studying human visual processing. Following research roles at Caltech, UCL, and University of Sussex as JST PRESTO researcher and Associate Professor, he founded Araya, Inc. in 2015. He researches consciousness principles in brains and machines through neuroscience-AI fusion, translating findings into industrial applications. Awards include the Young Scientist Award from Japan's Ministry of Education. Since 2020, he manages the Moonshot Project developing AI-assisted brain-machine interfaces. In 2024, he served as Vice Chair of UNESCO's Ad Hoc Expert Group drafting neurotechnology ethics recommendations. |
| **Prof. Yoshinori Yamakawa** is currently the Kyoto University Professor of Graduate School of Management and the chairman of Brain Impact, a general incorporated association responsible for the social development of the ImPACT program of the Cabinet Office in Japan. Through the Brain Healthcare Quotients (BHQ, H.861), a brain health management index approved to international standards, he promotes the improvement of people's wellbeing, mental health care for employees and dementia prevention for the elderly. For his activities in industry-government-academia collaboration, Professor Yamakawa was awarded the Japan Open Innovation Prize (JOIP) in 2024. |

**Dr. Chieko Asakawa** is an IBM Fellow, working in the area of accessibility. Since 2010, Chieko is focusing on real world accessibility to help the visually impaired to better comprehend their surroundings and navigate the world by the power of AI. Her latest project is the development of the AI suitcase, a navigation robot for the visually impaired. Dr. Asakawa concurrently serves as Chief Executive Director of the Japanese National Museum of Emerging Science and Innovation (Miraikan) since 2021. In 2013, the government of Japan awarded the Medal of Honor with Purple Ribbon to Dr. Asakawa for her contributions to accessibility research. She was elected as a foreign member of the US National Academy of Engineering in 2017, inducted into the National Inventors Hall of Fame (NIHF) in 2019.

**Prof. Didier Samuel** is a physician, researcher and hepatology specialist. Before his appointment as Chairman and CEO of Inserm on February 1, 2023 he was Dean of the Faculty of Medicine of the University of Paris-Saclay since 2017 and Director of the Hepatology and Hepatic Intensive Care Department and Liver Transplantation Program at the Paul-Brousse Hospital. Prof. SAMUEL chaired the Conference of Deans of Medicine and French National Research Coordination Committee. Since 2005, Prof. SAMUEL has headed the Inserm Research Unit devoted to physiopathogenesis and treatment of liver diseases. As an active Member of the International Society for Liver Transplantation between 2007 and 2015, his expertise in the field of liver diseases and liver transplantation is internationally recognized.

## References

Abdelraheem, O. M., Ikiz, B., Chun, S., Eyre, H., & Salama, M. (2025). Quantifying ecological intelligence: Building metrics for the green brain capital model—a systematic review. *BMJ Mental Health*, *28*(1). e301317. doi:https://doi.org/10.1136/bmjment-2024-301317

Baumann, A., Audibert, G., Lafaye, C. G., Puybasset, L., Mertes, P.-M., & Claudot, F. (2013). Elective non-therapeutic intensive care and the four principles of medical ethics. *Journal of Medical Ethics*, *39*(3), 139–142. doi:https://doi.org/10.1136/medethics-2012-100990

Bethlehem, R. A. I., Seidlitz, J., White, S. R., Vogel, J. W., Anderson, K. M., Adamson, C., Adler, S., Alexopoulos, G. S., Anagnostou, E., Areces-Gonzalez, A., Astle, D. E., Auyeung, B., Ayub, M., Bae, J., Ball, G., Baron-Cohen, S., Beare, R., Bedford, S. A., Benegal, V., … Alexander-Bloch, A. F. (2022). Brain charts for the human lifespan. *Nature*, *604*(7906), 525–533. doi:https://doi.org/10.1038/s41586-022-04554-y

Duffner, L. A., DeJong, N. R., Jansen, J. F. A., Backes, W. H., De Vugt, M., Deckers, K., & Köhler, S. (2023). Associations between social health factors, cognitive activity and neurostructural markers for brain health – A systematic literature review and meta-analysis. *Ageing Research Reviews*, *89*, 101986. doi:https://doi.org/10.1016/j.arr.2023.101986

Elkind, M. S. V., Eyre, H., & Pegus, C. (2025). An Advocacy Agenda for the Human Brain: Moving From Brain Health to Brain Capital. *Circulation*, *151*(3), e24–e26. https://doi.org/10.1161/CIRCULATIONAHA.124.073077

Erickson, K. I., Weinstein, A. M., & Lopez, O. L. (2012). Physical Activity, Brain Plasticity, and Alzheimer's Disease. *Archives of Medical Research*, *43*(8), 615–621. doi:https://doi.org/10.1016/j.arcmed.2012.09.008

Eyre, H. A., Hynes, W., Ayadi, R., Manes, F., & Swieboda, P. (2024). Brain capital is crucial for global sustainable development. *The Lancet Neurology*, *23*(3), 233–235. doi:https://doi.org/10.1016/S1474-4422(24)00031-0

Harrison, S. L., Sajjad, A., Bramer, W. M., Ikram, M. A., Tiemeier, H., & Stephan, B. C. M. (2015). Exploring strategies to operationalize cognitive reserve: A systematic review of reviews. *Journal of Clinical and Experimental Neuropsychology*, *37*(3), 253–264. doi:https://doi.org/10.1080/13803395.2014.1002759

Hauglund, N. L., Pavan, C., & Nedergaard, M. (2020). Cleaning the sleeping brain – the potential restorative function of the glymphatic system. *Current Opinion in Physiology, 15*, 1–6. doi:https://doi.org/10.1016/j.cophys.2019.10.020

Ibanez, A., & Eyre, H. (2023). Brain capital, ecological development and sustainable environments. *BMJ Mental Health, 26*(1), 1–3. doi:https://doi.org/10.1136/bmjment-2023-300803

Kolappa, K., Seeher, K., & Dua, T. (2022). Brain health as a global priority. *Journal of the Neurological Sciences, 439*. 120326. doi:https://doi.org/10.1016/j.jns.2022.120326

Liu, J., Li, Y., Zhao, D., Zhong, L., Wang, Y., Hao, M., & Ma, J. (2025). Efficacy and safety of brain–computer interface for stroke rehabilitation: An overview of systematic review. *Frontiers in Human Neuroscience, 19*, 1525293. doi:https://doi.org/10.3389/fnhum.2025.1525293

Miao, B. Y., Sushil, M., Xu, A., Wang, M., Arneson, D., Berkley, E., Subash, M., Vashisht, R., Rudrapatna, V., & Butte, A. J. (2024). Characterisation of digital therapeutic clinical trials: A systematic review with natural language processing. *The Lancet Digital Health, 6*(3), e222–e229. doi:https://doi.org/10.1016/S2589-7500(23)00244-3

Nail-Beatty, O., Ibanez, A., Ayadi, R., Swieboda, P., Njamnshi, A. K., Occhipinti, J.-A., Hynes, W., Ikiz, B., Castro-Aldrete, L., O'Brien, K., Platt, M. L., Adalat, S., Abdullah, J. M., Dhamija, R. K., Merali, Z., Mostert, C., Beck, D., Saxena, S., Salama, M., … Eyre, H. A. (2024). Brain health is essential for smooth economic transitions: Towards socio-economic sustainability, productivity and well-being. *Brain Communications, 6*(6), fcae360. doi:https://doi.org/10.1093/braincomms/fcae360

Reardon, S. (2016). 'Brain doping' may improve athletes' performance. *Nature, 531*(7594), 283–284. doi:https://doi.org/10.1038/nature.2016.19534

Segato, A., Marzullo, A., Calimeri, F., & De Momi, E. (2020). Artificial intelligence for brain diseases: A systematic review. *APL Bioengineering, 4*(4), 041503. doi:https://doi.org/10.1063/5.0011697

Shen, Q.-Q., Hou, J.-M., Xia, T., Zhang, J.-Y., Wang, D.-L., Yang, Y., Luo, R., Xin, Z.-L., Yin, H., & Cui, L. (2024). Exercise promotes brain health: A systematic review of fNIRS studies. *Frontiers in Psychology, 15*. 1327822 doi:https://doi.org/10.3389/fpsyg.2024.1327822

Smith, E., Ali, D., Wilkerson, B., Dawson, W. D., Sobowale, K., Reynolds, C., Berk, M., Lavretsky, H., Jeste, D., Ng, C. H., Soares, J. C., Aragam, G., Wainer, Z., Manji, H. K., Licinio, J., Lo, A. W., Storch, E., Fu, E., Leboyer, M., … Eyre, H. A. (2021). A Brain Capital Grand Strategy: Toward economic reimagination. *Molecular Psychiatry, 26*(1), 3–22. doi:https://doi.org/10.1038/s41380-020-00918-w

Tai, X. Y., Chen, C., Manohar, S., & Husain, M. (2022). Impact of sleep duration on executive function and brain structure. *Communications Biology, 5*(1), 201. doi:https://doi.org/10.1038/s42003-022-03123-3

Voss, M. W., Nagamatsu, L. S., Liu-Ambrose, T., & Kramer, A. F. (2011). Exercise, brain, and cognition across the life span. *Journal of Applied Physiology, 111*(5), 1505–1513. doi:https://doi.org/10.1152/japplphysiol.00210.2011

Wang, J., Chen, T., Xie, J., Zhao, S., Jiang, Y., Zhang, H., & Zhu, W. (2024). A bibliometric analysis of international publication trends in brain atrophy research (2008–2023). *Frontiers in Neurology, 15*. 1348778. doi:https://doi.org/10.3389/fneur.2024.1348778

Wang, Y., Pan, Y., & Li, H. (2020). What is brain health and why is it important? *BMJ, 371* m3683. doi:https://doi.org/10.1136/bmj.m3683

Zang, G. Y., Rao, K., Wu, J., He, Y., Tang, Y., & Shi, L. (2025). Digital Therapeutics for Cognitive Impairment: Exploring Innovations, Challenges, and Future Prospects. *Journal of Medical Internet Research, 27*(1), e73689. https://doi.org/10.2196/73689

# Part III
# Science, Business, and Policy for Innovation

# Chapter 8
# Basic Science, Innovation, and Policy

**Cristian Mejia**

## 1 Introduction

Science stands at a precarious crossroads. Never before has humanity possessed such powerful tools for discovery—artificial intelligence accelerates drug development, quantum computers promise to solve intractable problems, and global communication networks enable instantaneous knowledge sharing across continents. Yet, paradoxically, the geopolitical order that facilitated this unprecedented scientific integration is fragmenting. The tension between technological acceleration and political fragmentation defines our era, creating what researchers have termed the "stockyard paradox": while fencing in research to protect national interests may seem prudent, it risks weakening the very ecosystem that produces scientific breakthroughs (Serger & Martins, 2023).

At the heart of this paradox lies a fundamental question about the nature of scientific knowledge itself. As Nobel laureate Konstantin Novoselov articulated during the Basic Science, Innovation, and Policy panel at the 2025 Science and Technology in Society *forum*:

> Scientific results are never proprietary—they are always global. They belong to humanity, almost in a physical sense, like air, which we breathe here in Japan. Does it belong to Japan? It's a global phenomenon. Absolutely the same with scientific knowledge.—**Konstantin Novoselov**

This philosophical foundation—that scientific knowledge constitutes a global commons—increasingly clashes with national security imperatives and strategic competition for technological leadership.

The challenge is not merely rhetorical. Between 2018 and 2025, the OECD documented a tenfold increase in national research security policies, rising from under

C. Mejia (✉)
Institute for Future Initiatives, The University of Tokyo, Tokyo, Japan



H. Komiyama et al. (eds.), *AI, Innovation, and Beyond*,
https://doi.org/10.1007/978-981-95-7020-1_8

30 to over 250 measures globally (OECD, 2025). Export controls that once targeted specific military technologies now encompass quantum computing, artificial intelligence, and biotechnology—fields where the boundary between civilian and military applications has become impossibly blurred (U.S. Bureau of Industry and Security, 2024). Universities, traditionally spaces of open inquiry, now implement Technology Control Plans that partition laboratories, screen researchers by nationality, and restrict data sharing in ways that fundamentally conflict with the academic mission (MIT Research Administration, 2024).

The shift from "open by default" to "securitized science" reflects deeper geopolitical realignments. For three decades following the Cold War's end, international scientific collaboration flourished under the assumption that economic integration and knowledge sharing would drive political convergence. The data seemed to support this optimism: internationally coauthored papers more than doubled over 20 years, with collaboration networks becoming denser and more global (Wagner et al., 2015). China emerged as a scientific powerhouse, with US-China collaborative publications increasing from approximately 100 per year before 2017 to 796 during the Biden administration (Li et al., 2024).

> The gap between scientific discovery and real-world application is narrowing at a remarkable speed, and the integration of science and business is accelerating. The pace of change we now observe would have been unimaginable in the previous century.—**Toshiko Abe**

This acceleration paradox—where technological advancement speeds up while political constraints multiply—creates unprecedented pressures on the research enterprise.

The consequences manifest across multiple dimensions. Following Russia's 2022 invasion of Ukraine, Western nations terminated bilateral Arctic research programs, creating a massive "data hole" that undermines global climate models covering 50% of the Arctic coastline (Rees & Büntgen, 2024). The U.S. Department of Justice's "China Initiative," though formally ended, left a lasting climate of suspicion that drives reverse brain drain as Asian-American scientists relocate to opportunities outside the United States (Lee & Haupt, 2021). European nations navigate an uneasy middle path, attempting to "de-risk" relationships with China while maintaining research ties that have proven scientifically productive (Shih et al., 2023).

Perhaps most concerning is the impact on fundamental research itself. As governments prioritize mission-oriented funding for technologies with clear national security applications, support for curiosity-driven basic science erodes. The 2025 U.S. budget proposals indicated cuts to the National Institutes of Health by approximately 39% and the National Science Foundation by nearly 56%, even as defense R&D budgets remained robust (Association of American Universities, 2025). Federal R&D spending, which once emphasized exploration of fundamental questions, increasingly focuses on near-term development and technology transition. The economic consequences may be severe: analyses suggest that a 20% cut to federal R&D could reduce U.S. GDP by over $1 trillion cumulatively over a decade (Information Technology and Innovation Foundation, 2025).

These transformations raise fundamental questions about the future of science itself. Drawing from the framing articulated by Mitch Davies, President of the National Research Council Canada, during the STS *forum* session on basic science, innovation, and policy, this chapter examines three interconnected challenges:

First, how do we maintain fundamental discovery science in increasingly mission-oriented funding environments? The history of transformative breakthroughs—from the internet to mRNA vaccines—demonstrates that revolutionary innovations often emerge from research that lacks clear applications at the time of discovery.

> It was a fringe area, a fringe pursuit, which did get support. And of course, here we are talking about it as if it's something we always knew would come. But the real answer is we had no clue whatsoever.—**Mitch Davies**

How do research systems maintain space for such unpredictable discovery when political pressures demand tangible returns and strategic relevance?

Second, can international collaboration survive when trust erodes and security concerns mount? Scientific progress has always depended on the free exchange of ideas, the mobility of talented researchers, and the ability to build on work conducted anywhere in the world. Yet visa restrictions, export controls, and institutional screening mechanisms now impede precisely these collaborative activities. Do viable models exist for maintaining productive scientific partnerships across geopolitical divides, or must we accept a future of competing scientific blocs—Western, Chinese, and perhaps others—each pursuing parallel research programs with limited exchange?

Third, how do we support early-career researchers navigating unprecedented complexity? Young scientists entering the field today face contradictory pressures: publish openly yet protect sensitive information, collaborate internationally while avoiding "countries of concern," pursue curiosity while demonstrating strategic relevance, and build careers in an environment where funding volatility and administrative burdens increasingly displace time for actual research. What institutional structures and policy frameworks can preserve the conditions that allow emerging researchers to develop the deep expertise and creative freedom required for breakthrough science?

## 2 Research Approach

This chapter synthesizes multiple streams of evidence to address these questions. Our primary qualitative source consists of insights from the October 5, 2025 STS *forum* panel on "Basic Science, Innovation and Policy," which convened leaders spanning government policy (Minister Abe of Japan), national laboratory direction (Dr. Paul Kearns of Argonne National Laboratory), scientific academies (Dr. Bettina Rockenbach of the German National Academy of Sciences Leopoldina), research council leadership (Dr. Mitch Davies of Canada's National Research Council), and

Nobel Prize-winning research experience (Dr. Konstantin Novoselov). This diversity of perspectives—from those setting policy to those conducting research—provides grounded insights into how geopolitical tensions manifest in practice and what responses prove effective.

To complement these expert perspectives, we conducted a systematic literature review examining how international scientific collaboration has evolved under geopolitical pressure. Using semantic search across millions of academic papers from Semantic Scholar and OpenAlex, we identified empirical studies that examine collaboration dynamics across diverse contexts: from World War I-era disruptions to contemporary US-China tensions, from Cold War particle physics to present-day Arctic research. These studies employ quantitative bibliometric analysis, case study methodologies, policy analysis, and comparative frameworks to document collaboration patterns, identify mechanisms of resilience and fracture, and evaluate the effectiveness of different preservation strategies.

# 3 Discussion

## *3.1 The Mechanisms of Collaboration Persistence and Decline*

Examination of international scientific collaboration across diverse geopolitical contexts reveals that neither inevitable decoupling nor unaffected cooperation accurately describes the current reality. Instead, collaboration persists or declines based on specific, identifiable mechanisms that shape how researchers, institutions, and nations navigate political tensions. Four mechanisms emerge as particularly consequential.

### 3.1.1 Institutional Autonomy and Risk-Based Frameworks

> The 'who' is who's doing the research, who's collaborating in this space, understanding their affiliations. The 'what' is the science that they intend to do, understanding the risk associated with that from a variety of perspectives. And the 'where' is what facilities are appropriate to do the research, and can we assure ourselves it will be conducted in a safe and responsible way.—**Paul Kearns**

The structure of policy implementation profoundly affects collaboration outcomes. This framework, termed the "managed research environment," represents a discretionary, case-by-case approach that preserves researcher autonomy while addressing legitimate security concerns.

Comparative evidence strongly supports this approach. Shih et al.'s analysis of Swedish and Australian responses to US-China tensions demonstrates stark divergence in outcomes (Shih et al., 2023). Sweden implemented guideline-based "responsible internationalization" through funding agencies like STINT, emphasizing reflection and institutional judgment rather than top-down mandates. Co-authored publications between Sweden and China increased 215% from 2012 to 2021. Australia, conversely, enacted comprehensive legislation including the Defence Trade Controls Act and Foreign Relations Act, granting ministerial veto powers over research grants and establishing the University Foreign Interference Taskforce with sector-wide due diligence requirements. Despite this—or perhaps because of it—Australia-China publications grew 274% in the same period, though the more restrictive framework created greater compliance burdens and uncertainty for researchers.

The key distinction lies not in whether collaboration increased, but in how institutions navigate restrictions. Where policies emphasize discretion and risk management, institutions develop internal expertise and researchers maintain agency. Where rigid compliance frameworks dominate, collaboration becomes administratively costly and researchers self-censor to avoid bureaucratic entanglement. The JASON Advisory Group's, 2024 recommendation to the NSF advocated precisely this distinction: security protocols customized project-by-project in consultation with principal investigators, rather than blanket restrictions that stifle innovation while providing illusory security (JASON Advisory Group, 2024).

### 3.1.2 Crisis Urgency Overrides Political Constraints

Acute existential threats temporarily supersede geopolitical rivalry. The COVID-19 pandemic provided a natural experiment: as the virus spread globally in early 2020, international collaboration on COVID-19 research surged to 32% of publications compared to 26% in preceding years (Fry et al., 2020). Most strikingly, US-China collaboration on COVID-19 exceeded rates from the previous 5 years despite escalating trade tensions and technology restrictions (Lee & Haupt, 2021). The imperative to understand transmission dynamics, develop vaccines, and identify treatments overwhelmed political constraints, demonstrating that when threats are sufficiently urgent and universal, scientists and their institutions prioritize solving problems over navigating political restrictions.

However, the structure of crisis collaboration differed from routine research. Fry et al. found that COVID-19 research teams were smaller and included fewer nations than preceding international collaborations (Fry et al., 2020). Both the US and China increased collaboration with each other but partnered with fewer countries overall—77 for the US and 64 for China on COVID-19 research. This pattern favored elite structures and consolidated networks rather than broadening participation, suggesting that crisis mobilization operates through different channels than steady-state cooperation.

Mission-oriented funding driven by crisis imperatives proves effective for engineering challenges where the goal is clear. It proves far less effective for paradigm-shifting discoveries that emerge from undirected curiosity. Crisis urgency can overcome collaboration barriers temporarily, but it simultaneously narrows the aperture of inquiry precisely when breadth matters most for unanticipated breakthroughs.

### 3.1.3 Network Redundancy and Multilateral Pathways

When bilateral relationships face political constraints, multilateral partnerships create alternative collaboration routes. The Cuban-US case provides compelling evidence. Despite a comprehensive US embargo established in 1961 that prohibited exchanges and restricted funding, Cuba-US scientific collaboration sustained growth over 40 years (Ronda-Pupo, 2021). The Smithsonian Institution and Cuba's Academy of Sciences agreement in 1980 significantly increased collaborative papers. President Obama's 2015 normalization efforts accelerated this trend. By 2020, Cuba had expanded scientific links to 80% of the world's countries, creating multiple pathways for knowledge exchange that reduced dependence on any single bilateral relationship.

This network embeddedness operates through preferential attachment mechanisms: researchers embedded in large international networks with redundant connections maintain productivity even when specific bilateral ties face restrictions.

> We need an international science policy that is open-minded and realistic. It should follow the guiding principle: as open as possible, as closed as necessary.—**Bettina Rockenbach**

This formulation captures the essence of network redundancy—maintaining maximum connectivity while acknowledging that some specific connections may require restriction for legitimate security reasons.

Contemporary examples include the Trillion Parameter Consortium led by Argonne National Laboratory, RIKEN (Japan), and Barcelona Supercomputing Center (Spain), which has grown to over 80 organizations spanning government, academia, and industry (Trillion Parameter Consortium, 2024). This multilateral structure for AI development creates alternative collaboration pathways that no single bilateral tension can easily disrupt. Similarly, the Horizon Europe "association" model allows non-EU countries meeting democracy and intellectual property criteria to participate in research programs, creating a federation of open science nations that transcends traditional alliance structures (European Commission, 2024).

### 3.1.4 Trust Erosion Versus Gradual Adaptation

The speed and nature of political rupture determines whether institutional buffers can adapt. Russia-West Arctic research provides a stark case. Following Russia's February 2022 invasion of Ukraine, the "A7" nations (Canada, Denmark, Finland, Iceland, Norway, Sweden, United States) immediately paused Arctic Council activities. While project-level work resumed by June 2022 for projects not involving Russia (approximately 70 of 140 projects), diplomatic-level engagement remains frozen (Arctic Council, 2022). The exclusion of Russia, which covers approximately 50% of Arctic coastline, created a massive data gap undermining global climate models' ability to predict permafrost thaw, methane release, and vegetation dynamics (Rees & Büntgen, 2024).

This collapse occurred because the trust rupture happened too rapidly for institutional mechanisms to adapt. Rees and Büntgen document the shift from established collaboration to "suspicion and fear" in a timeframe measured in days, not years. Peer-to-peer connections that might have preserved some research continuity require time to develop and cannot substitute for suddenly severed institutional ties. The Arctic Council's consensus-based governance, lacking legal personality or binding treaty foundation, provided no resilience when political consensus evaporated.

Contrast this with US-China dynamics. Despite escalating tensions since 2017, some collaboration increased even as overall patterns shifted. Li et al. found that US-China collaborative publications on sustainable development reached 796 during the Biden administration, up from 537 during the Trump years and approximately 100 annually before 2017 (Li et al., 2024). This occurred despite—or perhaps strategically in response to—increasing restrictions in other domains. The gradual nature of US-China tension escalation allowed researchers and institutions to adapt: shifting to less sensitive topics, utilizing third-country collaborations, and developing new partnership models. Time enables institutional learning that acute rupture precludes.

## *3.2 Institutional Architectures of Resilience*

While mechanisms explain why collaboration persists or fails in specific contexts, institutional design determines whether organizations can weather geopolitical storms. Examination of major international scientific institutions reveals that governance structures matter enormously, with specific design features correlating strongly with resilience or fragility.

### 3.2.1 CERN: Scientific Meritocracy with Political Insulation

The European Organization for Nuclear Research (CERN) exemplifies resilient institutional architecture. Established in 1954 during Cold War tensions, CERN has survived multiple geopolitical crises while maintaining scientific productivity and expanding membership to 23 member states plus associate members and observers worldwide (CERN, 2024).

Three governance features explain this resilience. First, the Scientific Policy Committee evaluates research merit through peer election based on scientific eminence without reference to nationality. This structurally separates scientific evaluation from political decision-making—member state delegates in the Council may vote on budgets and administrative matters, but cannot override scientific judgment on research priorities or collaboration decisions. Second, CERN's budget derives from treaty obligations that insulate funding from short-term political fluctuations or austerity measures in individual member states. Third, CERN's convention explicitly prohibits military work, creating a "safe space" that builds trust among diverse members. Member states know research cannot be secretly weaponized, enabling Israeli and Iranian physicists, Russian and Ukrainian researchers, Chinese and American scientists to work side-by-side on fundamental questions about the universe's structure.

These neutral scientific spaces create value precisely because they transcend national boundaries while respecting national security concerns through structural prohibitions rather than case-by-case screening.

### 3.2.2 SESAME: Diplomacy in Conflict Zones

SESAME (Synchrotron-light for Experimental Science and Applications in the Middle East) in Jordan demonstrates that the CERN model can work even in the world's most volatile regions. Bringing together Cyprus, Egypt, Iran, Israel, Jordan, Pakistan, Palestine, and Turkey, SESAME operates a synchrotron light source where scientists from nations with no diplomatic relations collaborate on research ranging from archaeology to materials science (Paolucci et al., 2021).

The governance innovations adapt CERN principles to more challenging political context. Operating under UNESCO auspices provides diplomatic immunity and neutral international legal status. Jordan hosts the facility, but scientists work on "international ground" rather than Jordanian soil, similar to how embassies function. Decision-making on membership and budget requires two-thirds majority, but critically, member expulsion requires unanimous vote of all other members. This high bar protects controversial members like Iran or Israel from removal due to shifting regional alliances or temporary conflicts.

SESAME represents "bottom-up" science diplomacy driven by scientists themselves rather than imposed by foreign ministries. The shared infrastructure—member states contributed beamlines and equipment—creates mutual investment that survives political tensions. When relations between certain member states deteriorate, the multilateral structure provides alternative pathways: Iranian and Israeli scientists may not collaborate bilaterally on specific experiments, but both participate in the broader SESAME community and benefit from the facility's capabilities.

### 3.2.3 The Arctic Council: Failure of Consensus-Based Forums

The Arctic Council's fracture post-2022 illustrates how governance structures lacking legal foundations collapse under acute political stress. Established in 1996 to facilitate cooperation on Arctic environmental protection and sustainable development, the Council cultivated "Arctic exceptionalism"—a belief that the region's unique challenges would insulate it from geopolitical rivalries (Arctic Council, 2022).

This proved illusory. The Council operates as a forum rather than a treaty-based organization, lacking legal personality and depending entirely on political consensus. When Russia invaded Ukraine, the A7 nations immediately paused all activities. While project-level work partially resumed, the diplomatic foundation remains shattered. Russia now pursues Arctic research partnerships primarily with China, creating a bifurcation into “Western” and “Russian-Asian” spheres that fundamentally undermines the Council’s original purpose of circumpolar cooperation (Arctic Institute, 2024).

The consequences extend far beyond organizational dysfunction. Climate scientists cannot accurately model Arctic warming without Russian territorial data covering half the Arctic coastline. Rees and Büntgen document massive data gaps in permafrost monitoring, carbon flux measurements, and vegetation biomass predictions that degrade global climate projections (Rees & Büntgen, 2024). The Council’s consensus requirement—which enabled broad participation when relations were good—became a fatal vulnerability when consensus evaporated. No fallback mechanisms existed to preserve even minimal scientific exchange when political relations collapsed.

### 3.2.4 The International Space Station

The International Space Station demonstrates that deep technical interdependence can force cooperation even during war, but represents a legacy model unlikely to be replicated. The ISS survives because the U.S. segment provides power while the Russian segment provides propulsion—neither can function independently (NASA, 2023). However, Russia has committed only through 2028 while Western partners plan to 2030, signaling future decoupling toward separate space stations aligned with geopolitical blocs.

Key resilience and fragility factors of these institutions are shown in Table 8.1.

Three design principles emerge. First, legal personality and treaty foundations provide resilience against political shocks by establishing institutional continuity

**Table 8.1** Comparative analysis: Design principles for resilient institutions

| *Institution* | *Governance model* | *Status (2025–26)* | *Key resilience/fragility factors* |
|---|---|---|---|
| CERN | Treaty-based IGO with scientific merit system | Stable, thriving | Military prohibition creates safe space; peer-based scientific governance; budget insulation from political cycles |
| SESAME | UNESCO umbrella with supermajority protections | Stable, expanding | Unanimous expulsion requirement protects controversial members; diplomatic immunity; bottom-up scientist-driven model |
| Arctic Council | Consensus forum lacking legal personality | Fractured, paused | No treaty foundation; depends entirely on political consensus; no fallback mechanisms when consensus fails |
| ISS | Intergovernmental agreement with technical interdependence | Operational but fragile | Physical interdependence forces cooperation; US provides power, Russia provides propulsion; timeline divergence signals future decoupling |

independent of current political relations. Second, separation of scientific and political tracks through merit-based governance insulates research priorities from immediate political pressures while maintaining governmental oversight on budgets and administration. Third, structural transparency (such as military prohibitions) builds trust more reliably than case-by-case screening, because members know institutional rules prevent activities they fear rather than relying on current political relationships.

## *3.3 The Funding Paradox: Mission Versus Curiosity*

Geopolitical competition drives a fundamental shift in research funding priorities that threatens the long-term vitality of scientific discovery. Governments increasingly direct resources toward mission-oriented programs with clear strategic applications—artificial intelligence for military advantage, quantum computing for cryptographic supremacy, biotechnology for pandemic preparedness—while support for curiosity-driven fundamental research erodes. This creates a utilitarian trap: short-term strategic gains come at the expense of the unpredictable breakthroughs that drive long-term competitive advantage.

The evidence of this shift appears across major science-funding nations. The 2025 U.S. budget proposals indicated cuts to the National Institutes of Health by approximately 39% and the National Science Foundation by nearly 56% compared to previous baselines, even as defense R&D spending remained robust (Association of American Universities, 2025). The rebalancing reflects prioritization of development over research, near-term applications over fundamental questions. Federal R&D spending patterns show this trajectory: whereas government once funded basic and applied research at twice industrial levels, today these shares have converged, with federal spending increasingly concentrated on late-stage development and technology transition (Congressional Research Service, 2024).

This reorientation toward missions appears superficially pragmatic. The Manhattan Project and Apollo Program demonstrated that focused, goal-oriented research can achieve seemingly impossible objectives. However, these historical examples involved engineering challenges where the fundamental science was already understood—nuclear fission was discovered decades before Los Alamos, and orbital mechanics preceded Kennedy's moon commitment. Mission-oriented funding excels at translating known science into applications but proves far less effective for generating paradigm shifts.

> Let us recall the words of Abraham Flexner, a Leopoldina member, who famously described basic research as the 'usefulness of useless knowledge.' Even seemingly useless findings today may become the basis for transformative innovation tomorrow.—**Bettina Rockenbach**

This principle captures the fundamental tension: the most valuable discoveries often cannot be anticipated at the time of investment. The laser emerged from quantum mechanics research with no practical application in mind; GPS depends on Einstein's relativity, developed to understand space-time rather than navigation; mRNA vaccine platforms built on decades of basic RNA biology funded for its intrinsic interest rather than pandemic preparedness.

The fusion energy research sector illustrates this dynamic starkly. The publicly funded International Thermonuclear Experimental Reactor (ITER) project maintains open science principles with equal intellectual property sharing among members—Japan, EU, USA, Russia, China, India, and South Korea collaborate despite geopolitical tensions because the framework ensures no single nation captures disproportionate advantage (Carayannis et al., 2024). Contrast this with the emerging DEMO phase, where private funding drives a shift toward proprietary technologies and exclusive possession. This transition from communal to competitive models may accelerate near-term commercialization but risks fragmenting the knowledge base that enabled ITER's progress in the first place.

The economic argument for prioritizing applied research over basic science proves fundamentally flawed when examined over meaningful timeframes. The Information Technology and Innovation Foundation estimates that a 20% reduction in federal R&D spending would reduce U.S. GDP by over $1 trillion cumulatively over a decade (Information Technology and Innovation Foundation, 2025). This occurs because basic research generates entirely new industries rather than incrementally improving existing ones—the economic multiplier effect of fundamental discovery far exceeds that of applied development.

> In our age of unprecedented uncertainty, where predicting the future has become increasingly difficult, it is the diversity and depth of our research ecosystem, along with the cultivation of highly skilled talent, that will determine our long-term social and economic vitality and, ultimately, our national strength.—**Toshiko Abe**

This framing connects research investment directly to national competitiveness, but crucially emphasizes ecosystem diversity rather than targeted programs. A robust research system maintains capacity across multiple fields, enabling rapid response when unanticipated challenges emerge or opportunities arise.

Funding concentration creates additional vulnerabilities through what sociologists term the "Matthew Effect"—advantages accumulate to already-advantaged actors. When basic research budgets contract while applied programs remain stable, well-funded elite institutions with defense contracts and large endowments survive while smaller institutions and younger researchers face elimination. This reduces intellectual diversity precisely when breadth of perspective matters most for breakthrough discovery. Analysis of Nobel Prize-winning research shows that truly transformative work often comes from unexpected sources, not necessarily the most prestigious or best-funded labs (Liang et al., 2023).

The historical record provides sobering warnings. Barriers to international cooperation demonstrably slow the production of basic science and its application in new technologies (Wagner et al., 2015). Internationally coauthored papers receive higher citations than single-nation papers, and the global collaboration network attracts and rewards highly productive researchers (Okamura, 2023). The World War I era offers particularly stark evidence: isolation led to dramatic declines in research deemed worthy of Nobel Prize nominations, with scientific progress stalling across the Allied-Central divide (Iaria et al., 2018).

Historically, universities relied on the Fundamental Research Exclusion (FRE) to exempt basic research from export controls. The FRE applies when resulting information is ordinarily published and shared broadly—the defining characteristic of academic science. However, the blurring boundary between basic and applied research in fields like quantum computing and artificial intelligence erodes this protection. When a quantum error correction algorithm might improve both medical imaging and missile guidance, or when machine learning techniques apply equally to drug discovery and autonomous weapons, the FRE's conceptual foundation weakens (University Research Compliance, 2024).

Funding agencies compound this erosion by inserting restrictions on publication or foreign national participation directly into grant contracts, effectively nullifying the FRE for those projects. Universities increasingly bifurcate into "open" fundamental research zones and "restricted" secure zones with Technology Control Plans requiring locked laboratories, badged access, personnel screening, and air-gapped computing (MIT Research Administration, 2024). This segregation discourages the serendipitous cross-pollination that defines academic innovation—a physicist working on a secure quantum project cannot casually discuss mathematical problems with a colleague who happens to be a foreign national.

> We cannot unknow; we cannot stop discovering. It is just part of human nature… I often laugh about those attempts that say we need to slow down the development of this or that technology. It's absolutely impossible.—**Konstantin Novoselov**

This observation highlights a fundamental asymmetry: knowledge, once created, diffuses. Attempting to contain it through restrictions may slow dissemination but cannot prevent eventual spread. The effort to fence off certain research areas from international collaboration thus imposes costs—reduced talent pools, slower iteration, duplicated effort—while providing only temporary protection against knowledge transfer.

The consequence is strategic disadvantage for nations imposing the strictest controls. If Country A restricts quantum research to citizens only while Country B maintains openness to global talent, Country B attracts superior researchers and advances faster. The "stockyard paradox" operates at national scale: high fences around research starve the innovation ecosystem they aim to protect (Serger & Martins, 2023).

## *3.4 The Human Capital Crisis*

Scientific progress ultimately depends on human talent—the researchers who generate insights, the graduate students who execute experiments, the postdocs who bridge discovery and application. Geopolitical tensions manifest most immediately through restrictions on human mobility: visa policies, screening mechanisms, and the chilling effects of security scrutiny. These barriers to talent flow create immediate damage to research productivity while generating long-term competitive disadvantage for nations that erect them.

### 3.4.1 Visa Wars and Mobility Restrictions

The United States, which has historically relied on foreign talent to power graduate programs and laboratories, faces self-inflicted wounds from increasingly restrictive immigration policies. Proposals to cap F-1 student and J-1 researcher visas at 4 years create fundamental incompatibility with doctoral timelines, which typically require 5–6 years (NAFSA, 2020). Students facing uncertainty about whether extensions will be granted rationally choose universities in Canada, the United Kingdom, Singapore, or other nations offering more predictable pathways.

The targeting proves particularly damaging. Policies specifically restrict students from "countries of concern"—primarily China and Iran—studying "sensitive" subjects including artificial intelligence, quantum computing, and advanced materials. This creates a presumption of guilt that deters top talent. A brilliant undergraduate in Shanghai considering PhD programs must weigh scientific opportunity against the risk of visa denial, sudden deportation, or investigation under programs like the discontinued but spiritually continuing China Initiative (Lee & Li, 2023). The rational response is to choose programs elsewhere, draining precisely the talent that drives American research competitiveness.

The quantitative impact appears in graduate enrollment data. Surveys of physics departments indicate 13–25% anticipated declines in international enrollment due to funding uncertainties and visa hostility (American Institute of Physics, 2023). Physics serves as a bellwether for STEM fields generally, suggesting a broader talent crisis. Chinese students in the U.S. declined nearly 22% while American students in China plummeted to 1.8% of 2018–2019 levels, reflecting deteriorating academic exchange in both directions (Institute of International Education, 2023).

> I think we put a lot of burden on them. In the organizations they get into, certainly in our environment too, we can ask a lot of them, and perhaps they're not spending their time on the areas they want to work in and being actually free to think and to have room to create.—
> **Mitch Davies**

This comment, made in the context of supporting early-career researchers, applies with particular force to navigating visa requirements, export control compliance, and security screening. Administrative overhead displaces research time, while uncertainty about collaborators' status disrupts the sustained relationships that enable ambitious projects.

The resulting "chilling effect" extends beyond direct restrictions. Asian-American scientists report feeling targeted, leading to reverse brain drain as established researchers relocate to opportunities in China, Singapore, or other nations offering less hostile environments (Lee & Li, 2023). This represents loss not only of individual talent but of networks, mentorship relationships, and institutional knowledge that took decades to build. High-profile scientists departing U.S. institutions take with them expertise and connections that damage American research capacity more than any foreign espionage could achieve.

### 3.4.2 Tiered Systems: Trusted Partners Versus the Rest

Even as barriers rise for some nations, they lower for others. AUKUS Pillar 2 includes provisions for streamlined researcher movement between the United States, United Kingdom, and Australia to facilitate cooperation on hypersonics, quantum technologies, and artificial intelligence (CSIS, 2023). This creates a "friction-free zone" for innovation among allies, enabling rapid iteration, resource pooling, and seamless collaboration.

The strategic logic appears sound: deep integration with trusted partners accelerates progress while mitigating security risks. However, this tiered approach risks creating echo chambers. If the next breakthrough in battery chemistry emerges from Vietnam, or if a Brazilian laboratory develops novel approaches to computational biology, a closed AUKUS loop might miss these advances until competitors incorporate them. Scientific leadership requires not only advancing the frontier but also monitoring the global landscape for innovations arising anywhere.

The tiered system also creates perverse incentives. Researchers from "trusted" nations face fewer barriers regardless of individual circumstances, while those from "countries of concern" face suspicion regardless of personal merit or research focus. This substitutes crude nationality screening for the granular risk assessment that actual security requires. A researcher from an allied nation studying weapons-applicable technology may face less scrutiny than a researcher from a rival nation studying astrophysics, inverting the sensible risk hierarchy.

## *3.5 Preservation Strategies: Technical and Structural Solutions*

The challenges documented above—eroding collaboration, mission-driven funding squeezing curiosity research, and mobility restrictions—demand innovative responses. If political trust has degraded, we must substitute technical trust. If bilateral relationships face constraints, we must build multilateral alternatives. If blanket restrictions create more damage than protection, we must develop granular risk frameworks. Evidence from successful institutional models and emerging technical capabilities suggests viable pathways for preserving collaborative science even amid geopolitical competition.

### 3.5.1 Calibrate Controls to Project-Level Risk

The JASON Advisory Group's, 2024 recommendations to the National Science Foundation represent a paradigm shift from categorical controls toward graduated risk assessment (JASON Advisory Group, 2024). Rather than applying uniform restrictions to broad technology categories, this approach classifies research projects along a risk spectrum and tailors protections accordingly.

A three-tier framework emerges from analysis of dual-use concerns and collaboration patterns. Low-risk research—fundamental physics, astronomy, pure mathematics, most environmental science—poses minimal security concerns because applications remain distant or indeterminate. Policy prescription: complete openness with standard academic norms. Medium-risk research—artificial intelligence safety, industrial biotechnology, advanced materials, civilian energy—has potential dual-use applications but substantial civilian benefits. Policy prescription: managed openness using technical safeguards like data enclaves, with collaboration decisions based on specific project characteristics rather than blanket prohibitions. High-risk research—weapons design, pathogen engineering with pandemic potential, cryptographic systems—presents clear security threats. Policy prescription: strict classification or export control with limited international collaboration restricted to trusted partners.

The critical innovation involves recognizing that most research falls into the medium category, where nuanced approaches prove superior to either complete openness or total restriction. The Swedish model demonstrates this empirically: guideline-based frameworks emphasizing institutional discretion and researcher reflection generated 215% growth in China collaboration while maintaining security awareness (Shih et al., 2023). Australia's rigid legislative approach generated even higher collaboration growth (274%) but with greater administrative burden and researcher uncertainty. The goal is not minimizing collaboration but optimizing the security-productivity tradeoff through flexible, context-sensitive policies.

The NSF's proposed "security by design" approach operationalizes this principle by engaging principal investigators in security planning before peer review rather than imposing restrictions afterward (JASON Advisory Group, 2024). This collaborative model builds researcher buy-in while developing custom protections matched to specific project risks. When scientists understand security rationales and participate in designing safeguards, compliance improves and innovation suffers less than under imposed mandates.

However, implementation challenges remain formidable. The "small yard, high fence" strategy risks expanding the yard as threat perceptions grow and technology boundaries blur. Quantum computing provides a cautionary example: initial export controls targeted specific devices, but subsequent expansions encompassed algorithms, materials, and even theoretical research that might eventually enable quantum advantage (U.S. Bureau of Industry and Security, 2024). Vigilance is required to prevent risk-based frameworks from devolving into categorical restrictions through incremental expansion.

### 3.5.2 Substitute Technical Trust for Political Trust

Technical solutions can substitute for political trust when properly deployed. Confidential computing and Trusted Execution Environments (TEEs) allow data processing in encrypted states, enabling researchers from rival nations to collaborate on sensitive datasets without exposing raw information (Russinovich et al., 2021). A Chinese researcher could execute algorithms on American healthcare data without seeing individual records, while American hosts cannot access the researcher's proprietary code. Computation occurs inside hardware-protected enclaves that maintain confidentiality for both parties.

The National COVID Cohort Collaborative (N3C) demonstrated this approach at scale, enabling global research on patient data without violating privacy or data sovereignty constraints (Haendel et al., 2021). Researchers from dozens of countries analyzed clinical outcomes, treatment effectiveness, and transmission patterns using a federated model where computations ran locally at each data site with only aggregated results shared. This architecture proves particularly valuable for dual-use research where methodology matters but data cannot cross borders.

For artificial intelligence collaboration, data enclaves enable joint safety testing without model weight exfiltration. Research teams can evaluate AI system behavior, probe for dangerous capabilities, and develop mitigation strategies within secure environments that prevent theft of intellectual property or training data. This addresses a critical vulnerability in AI governance: open collaboration on safety research requires sharing models, but model sharing enables competitors to capture economic value or military advantage. Data enclaves thread this needle by enabling evaluation without full disclosure.

Federated learning extends these principles further, allowing model training across distributed datasets that never centralize (Yang et al., 2019). A global consortium developing medical AI could train on American, European, Chinese, and Indian patient data without any single institution accessing the complete dataset. The model learns from all sources while data sovereignty remains intact. This proves essential for applications requiring diverse training data—global disease surveillance, climate modeling, genomic research—where comprehensive datasets are geographically dispersed and politically restricted.

Blockchain-based provenance systems offer complementary capabilities by creating tamper-evident records of data access, computation, and results (Al-Mamun et al., 2021). When researchers from rival nations collaborate in sensitive domains, immutable audit trails build confidence that neither party violated agreed constraints. For fields like nuclear verification or pandemic monitoring, where trust is low but verification needs are high, cryptographic assurance substitutes for political confidence.

### 3.5.3 Establish Neutral Zones and Safe Harbors

Physical and legal spaces insulated from national jurisdiction provide venues where constrained collaboration can proceed. The Marine Scientific Research (MSR) framework under the Law of the Sea Convention offers a precedent: research in Exclusive Economic Zones requires coastal state consent, but the Convention establishes norms facilitating scientific access even when political relations are poor (Intergovernmental Oceanographic Commission, 2024). This model could extend to digital domains through "Cyber-EEZs" or data havens where international treaties mandate scientific data sharing for global challenges like climate modeling (Rothwell & Scott, 2023).

Third-country hosting leverages neutral jurisdictions to facilitate sensitive collaborations. Switzerland and Singapore, nations with strong scientific infrastructure and diplomatic neutrality, increasingly host research projects involving rivals. The Geneva Science Diplomacy model proposes expanding this concept: international scientific organizations headquartered in neutral nations, governed by treaties that insulate scientific work from bilateral political tensions (Geneva Science and Diplomacy Anticipator, 2023). American and Chinese researchers might struggle to collaborate in either's homeland but can partner effectively on Swiss soil under Swiss legal jurisdiction.

SESAME demonstrates this principle's viability: Iranian and Israeli scientists collaborate in Jordan because the facility operates as international ground under UNESCO auspices (Paolucci et al., 2021). The legal architecture—diplomatic immunity, neutral governance, multilateral oversight—creates space for cooperation that would be impossible bilaterally. Expansion of this model could enable collaboration in other conflict zones or sensitive research domains.

International scientific platforms in space represent an untapped opportunity. The International Space Station showed how shared infrastructure creates mutual dependence that survives earthly conflicts, though its legacy design limits future applicability (NASA, 2023). Future neutral orbital research stations, perhaps governed by UN frameworks or new international treaties, could provide venues for genuinely global science insulated from terrestrial geopolitics. Research on space-based solar power, asteroid mining, or fundamental physics conducted in orbit under international law escapes national export controls while addressing challenges affecting all humanity.

### 3.5.4 Build Multilateral Alternatives to Bilateral Ties

When bilateral relationships face constraints, multilateral partnerships create alternative pathways. The Trillion Parameter Consortium exemplifies this approach: led by Argonne National Laboratory (USA), RIKEN (Japan), and Barcelona Supercomputing Center (Spain), with over 80 participating organizations spanning government, academia, and industry across multiple continents (Trillion Parameter Consortium, 2024). This network structure for AI development provides redundancy—no single bilateral tension can disrupt the entire collaboration. If US-China tensions restrict direct partnership, both can participate through Japanese or European intermediaries.

The Horizon Europe "association" model extends this logic to funding frameworks (European Commission, 2024). Non-EU countries meeting criteria for democracy, rule of law, and intellectual property protection can join the research program as associated members. Recent expansions to include New Zealand and Canada demonstrate the model's scalability beyond geographic proximity. This creates a "federation" of open science nations united by shared values and practices rather than traditional alliances. Importantly, participation requirements incentivize good behavior: nations must maintain academic freedom and research integrity standards to retain association status, creating soft power mechanisms that support scientific norms.

Regional collaborations offer additional resilience. The Asia-Pacific region hosts numerous multilateral science initiatives—APEC research programs, ASEAN scientific networks, Pacific Rim collaborations—that continue functioning even as specific bilateral relationships fluctuate. These overlapping networks create a mesh topology where multiple paths connect any two nodes, so blocking one connection merely shifts traffic through alternative routes.

Open-source consortia demonstrate the power of voluntary, self-organizing multilateral structures. Major scientific computing frameworks, genomic databases, and materials science repositories operate as community resources governed by contributor coalitions rather than single nations. When participation is voluntary and governance is decentralized, no single political authority can shut down the collaboration. This federated approach proves particularly effective for digital resources where geographic location matters less than network effects and community engagement.

### 3.5.5 Sustain Public Funding to Preserve Open Science

The ITER fusion energy project demonstrates that public funding structures enable sustained international collaboration where private funding would drive fragmentation (Carayannis et al., 2024). Equal intellectual property sharing among members—regardless of contribution size—ensures no single nation captures disproportionate advantage, maintaining incentives for even competitors to collaborate. When Japan, EU, USA, Russia, China, India, and South Korea all benefit equally from breakthroughs, all maintain commitment despite geopolitical tensions.

Contrast this with emerging private fusion ventures, where proprietary models and exclusive IP threaten to fracture the knowledge commons that enabled ITER's progress. Private funding accelerates development by mobilizing capital and focusing resources, but the shift from communal to competitive frameworks risks duplicating effort, fragmenting talent pools, and restricting knowledge flows. The optimal balance preserves public investment in fundamental research establishing shared knowledge foundations while enabling private sector innovation in applications and commercialization.

Evidence suggests that publicly funded research maintains higher rates of international collaboration than privately funded work across multiple fields. Government funding typically includes explicit openness requirements, international collaboration incentives, and data-sharing mandates. Private funding, especially from defense contractors or strategic industries, more often includes restrictions on foreign participation, publication delays, and proprietary data controls. This divergence means that declining government R&D budgets not only reduce total research investment but specifically reduce the open, collaborative research most valuable for global challenges.

Policy implications are clear: sustained public investment in fundamental research serves strategic interests precisely because it maintains open collaboration that attracts global talent and maximizes knowledge creation. The false economy of cutting basic research to fund applied development sacrifices long-term advantage for short-term applications. Nations that maintain robust public science funding while others contract will capture disproportionate shares of international collaboration, with compounding advantages as talent and partnerships concentrate in more open systems.

## 4 Conclusion

The evidence documented here—from World War I's collapse of international physics to today's fragmenting AI research landscape—demonstrates that geopolitical competition consistently threatens scientific collaboration, yet strategic choices determine whether competition destroys or merely strains the global research system. The current moment presents particular danger because technological acceleration in AI, biotechnology, and quantum computing creates both security imperatives driving restrictions and global challenges demanding collaboration. Managing this tension requires moving beyond false dichotomies of "open versus secure" toward nuanced frameworks that protect genuine security interests while preserving collaborative capacity.

Three strategic imperatives emerge from this analysis. First, precision over breadth in security measures. Risk-based frameworks distinguishing fundamental research from genuine dual-use activities prevent the expansive creep that transforms narrow protections into broad chilling effects. The Swedish guideline-based model demonstrates that managed openness maintains security awareness while enabling collaboration growth, whereas blanket restrictions generate compliance costs that discourage international engagement without commensurate security benefits. Second, technical substitutes for political trust. Confidential computing, data enclaves, and federated learning enable collaboration on sensitive research without requiring the political confidence that geopolitical competition erodes. When researchers from rival nations can jointly analyze data or test AI systems within architectures that maintain confidentiality, technical assurance substitutes for diplomatic trust. Third, multilateral resilience against bilateral tensions. Network structures where many paths connect any two nodes ensure that individual relationship failures cannot fragment the entire system. The Trillion Parameter Consortium's distributed governance and Horizon Europe's association model demonstrate how institutional design can build redundancy into collaborative frameworks.

For policymakers, the imperative is resisting reflexive restriction. Export controls and collaboration screening serve legitimate purposes, but each expansion carries accumulating costs in reduced knowledge flows, talent flight, and scientific competitiveness. The long-term strategic advantage lies in maintaining open systems that attract global talent and maximize discovery, not in fragmenting into competing blocs with reduced collective capability. For research institutions, the challenge involves building infrastructure that enables secure collaboration—data enclaves, technology control plans, compliance systems that protect researchers rather than merely institutional liability. For individual scientists, preservation requires both technical adaptation and institutional advocacy: learning to work within managed environments while pushing back against restrictions that serve bureaucratic risk-aversion rather than security imperatives.

The path forward follows Rockenbach's principle: "as open as possible, as closed as necessary". Determining necessity requires honest assessment of risks, not worst-case speculation. It requires distinguishing genuine security threats from competitive anxieties or institutional convenience. Most fundamentally, it requires recognizing that long-term security depends on scientific vitality, which depends on global collaboration. Nations that fracture the knowledge commons to capture temporary advantage will find themselves isolated from the discoveries that drive both prosperity and security. The challenge of this era is maintaining enough openness to preserve science's collaborative power while implementing enough protection to sustain political support for openness. The evidence suggests this balance remains achievable—but only through deliberate institutional design, technical innovation, and sustained commitment to collaboration's strategic value.

**Competing Interests** No competing interests.

## Session Panelists

**Davies, Mitch (Chair)**
President, National Research Council Canada (NRC), Canada
Mitch Davies returned to the NRC as President for a 5-year term on January 3, 2024, having previously served in the role from December 2020 to October 2021. Previously, he was Senior Assistant Deputy Minister, Strategic Innovation Fund, at Innovation, Science and Economic Development Canada (ISED), which provides major investments for innovation and industrial projects to grow Canada's economy. Since 2006, he has held a number of increasingly senior executive positions in science, innovation and industry policy in Canada's federal government, including Canada's Privy Council Office. Mr. Davies holds a Master of Business Administration degree from Queen's University and a Bachelor of Commerce (Honours) degree from the University of Manitoba.

**Abe, Toshiko**
Minister, Ministry of Education, Culture, Sports, Science and Technology (MEXT), Japan
Past Posts: House: 2011 Director, Special Committee on Children and Youth Affairs 2013 Director, Standing Committee on Health, Labour and Welfare 2015 Director, Standing Committee on Agriculture, Forestry and Fisheries 2016 Director, Special Committee on Political Ethics and Election Law 2017 Director, Special Committee on Political Ethics and Election Law 2017 Director, Standing Committee on Financial Affairs Administration: 2012 Parliamentary Secretary (now Parliamentary Vice-Minister) for Foreign 2014 Affairs State Minister of Agriculture, Forestry and Fisheries 2018 State Minister, Ministry for Foreign Affairs 2023 State Minister, Education, Culture, Sports, Science and Technology 2024 Minister, Education, Culture, Sports, Science and Technology

**Kearns, Paul K.**
Director, Argonne National Laboratory, U.S.A.
Paul Kearns is the Director of the U.S. Department of Energy's Argonne National Laboratory. A multidisciplinary science and engineering research center with a $1.1 billion diversified research portfolio. He is responsible for guiding development and implementation of the laboratory's strategic vision to deliver pivotal discoveries, realize dynamic collaborations, and provide powerful scientific tools and facilities. He establishes policies that make the laboratory more welcoming, inclusive, and supportive of its world-class community of talent. He oversees six DOE user facilities, six multi-institution collaborations, and many research programs that are critical to Argonne's mission of accelerating science and technology to drive U.S. prosperity and security.

**Novoselov, Konstantin Sergeevich**
Tan Chin Tuan Centennial Professor, Institute for Functional Intelligent Materials, National University of Singapore, Singapore; Professor of Condensed Matter Physics, The University of Manchester, U.K. [Nobel Laureate 2010 (Physics)]
Kostya Novoselov received his PhD from the University of Nijmegen, the Netherlands, before joining Andre Geim at the University of Manchester, UK for his post-doctoral training. In 2010, he and Andre Geim jointly received the Nobel Prize for Physics "for groundbreaking experiments regarding the two-dimensional material graphene". In 2019, he joined the National University of Singapore as Distinguished Professor and in 2021 was appointed Director of the newly established Institute for Functional Intelligent Materials in Singapore. His research spans across condensed matter physics; mesoscopic transport; superconductivity and ferromagnetism; nanostructures and nanofabrication; graphene and other two-dimensional crystals.

**Rockenbach, Bettina**
President, German National Academy of Sciences Leopoldina, Germany
The economist Bettina Rockenbach is an expert in the fields of Behavioral Economics and Game Theory. Since March 2025, she is president of the German National Academy of Sciences Leopoldina. She studied mathematics, economics and computer science at the University of Bonn, received her PhD in economics under the supervision of Nobel Laureate Reinhard Selten and completed her habilitation at the University of Bonn. She has been Professor of Microeconomics at the University of Erfurt, Professor of Experimental and Behavioral Economics at the University of Cologne, Director of the Reinhard Selten Institute at the Universities of Bonn and Cologne and Fellow of the Max Planck Institute for Research on Collective Goods. She held visiting positions at Caltech, Tilburg and San Diego.

## References

Al-Mamun, A., Feng, Y., & Dongfang, Z. (2021). SciChain: Blockchain-enabled lightweight and efficient data provenance for reproducible scientific computing. In *2021 IEEE 37th International Conference on Data Engineering (ICDE)*, 1775–1786. https://doi.org/10.1109/ICDE51399.2021.00166

American Institute of Physics. (2023). *US physics departments expect to shrink graduate programs*. https://www.aip.org/fyi/us-physics-departments-expect-to-shrink-graduate-programs

Arctic Council. (2022). *Arctic council advances resumption of project-level work*. https://arctic-council.org/news/arctic-council-advances-resumption-of-project-level-work/

Arctic Institute. (2024). *The arctic council in the shadow of geopolitics*. https://www.thearcticinstitute.org/arctic-council-shadow-geopolitics/

Association of American Universities. (2025). *Federal research cuts threaten U.S. innovation and leadership*. https://www.aau.edu/key-issues/federal-research-cuts-threaten-us-innovation-and-leadership.

Carayannis, Elias G., Vincenzi M/, and John Draper. (2024). The Economic Logic of Open Science in Fusion Energy Research: A Policy Perspective. *Energy Policy* 186, 113983. doi:https://doi.org/10.1016/j.enpol.2024.113983.

CERN. (2024). *Our governance*. https://home.cern/about/who-we-are/our-governance

Congressional Research Service. (2024). *Federal research and development (r&d) funding: FY2024*. https://crsreports.congress.gov/product/pdf/R/R47564

CSIS. (2023). *AUKUS pillar two: Advancing the capabilities of the United States, United Kingdom, and Australia*. https://www.csis.org/analysis/aukus-pillar-two-advancing-capabilities-united-states-united-kingdom-and-australia

European Commission. (2024). *Association to horizon Europe*. https://research-and-innovation.ec.europa.eu/strategy/strategy-research-and-innovation/europe-world/international-cooperation/association-horizon-europe_en

Fry, C., Cai, X., Zhang, Y., & Wagner, C. (2020). *Consolidation in a crisis: Patterns of international collaboration in COVID-19 research*. https://doi.org/10.2139/ssrn.3595455

Geneva Science and Diplomacy Anticipator. (2023*). GESDA 2023 science breakthrough radar*. GESDA. https://gesda.global/science-breakthrough-radar/

Haendel, Melissa A., Christopher G. Chute, Tellen D. Bennett, et al. 2021. The National COVID Cohort Collaborative (N3C): Rationale, Design, Infrastructure, and Deployment. *Journal of the American Medical Informatics Association* 28 (3): 427–43. doi:https://doi.org/10.1093/jamia/ocaa196.

Iaria, Alessandro, Carlo Schwarz, and Fabian Waldinger. 2018. Frontier Knowledge and Scientific Production: Evidence from the Collapse of International Science. *Quarterly Journal of Economics* 133 (2): 927–91. doi:https://doi.org/10.1093/qje/qjx046.

InformationTechnologyandInnovationFoundation.(2025).*HowreducingfederalR&DreducesGDP growth*. https://itif.org/publications/2025/09/15/how-reducing-federal-rd-reduces-gdp-growth/

Institute of International Education. (2023). *International student enrollment trends*. https://www.iie.org/research-and-insights/open-doors/

Intergovernmental Oceanographic Commission. (2024). *Marine scientific research under the UN convention on the law of the Sea*. https://www.unesco.org/msr-unclos

JASON Advisory Group. (2024). *Safeguarding the research enterprise*. https://www.nsf.gov/news/nsf-announcement-jason-report-safeguarding

Lee, Jenny J., and John P. Haupt. 2021. Scientific Collaboration on COVID-19 Amidst Geopolitical Tensions Between the US and China. *The Journal of Higher Education* 92 (2): 303–329. doi:https://doi.org/10.1080/00221546.2020.1827924.

Lee, Jenny, and Xiaojie Li. 2023. Neo-Racism, Neo-Nationalism, and the Costs for Scientific Competitiveness: The China Initiative in the United States. *Review of Higher Education* 46 (2), 285–309. doi:https://doi.org/10.1353/rhe.2023.0000.

Li, Rongrong, Feng Ren, and Qiang Wang. 2024. China-US Scientific Collaboration on Sustainable Development Amidst Geopolitical Tensions. *Humanities and Social Sciences Communications* 11, 1448. doi:https://doi.org/10.1057/s41599-024-03948-9.

Liang, Guoqiang, Yaxin Li, Liying Song, and Chaoguang Huo. 2023. Magnitude Decrease of the Matthew Effect in Citations: A Study Based on Nobel Prize Articles. *Scientometrics* 128 (11): 6063–6079. doi:https://doi.org/10.1007/s11192-023-04874-4.

MIT Research Administration. (2024). *Using restricted material at MIT.* https://research.mit.edu/security-integrity-and-compliance/export-control/scholarly-activities/using-restricted-material-mit

NAFSA. (2020). *DHS proposal to replace duration of status.* https://www.nafsa.org/regulatory-information/dhs-proposal-replace-duration-status

NASA. (2023). *Partners extend international space station for benefit of humanity.* https://www.nasa.gov/blogs/spacestation/2023/04/27/partners-extend-international-space-station-for-benefit-of-humanity/

OECD. (2025). *What Is Research Security and Why Does It Matter for Global Science?* https://www.oecd.org/en/blogs/2025/11/what-is-research-security-and-why-does-it-matter-for-global-science.html

Okamura, Keisuke. 2023. A Half-Century of Global Collaboration in Science and the "Shrinking World". *Quantitative Science Studies* 4 (4): 938–959. doi:https://doi.org/10.1162/qss_a_00268.

Paolucci, G. et al. 2021. SESAME—a Synchrotron Light Source in the Middle East: An Example of International Cooperation. *Open Research Europe* 1: 51. https://doi.org/10.12688/openreseurope.13762.1

Rees, Gareth, and Ulf Büntgen. 2024. Russian Dilemma for Global Arctic Science. *Ambio*.53 (8): 1246–1250 doi:https://doi.org/10.1007/s13280-024-02038-z.

Ronda-Pupo, Guillermo Armando. 2021. Cuba-u.s. Scientific Collaboration: Beyond the Embargo. *PLoS ONE* 16 (7), e0255106. doi:https://doi.org/10.1371/journal.pone.0255106.

Rothwell, Donald R., and Karen N. Scott. 2023. *The International Law of the Sea.* 3rd ed.. Oxford: Hart Publishing.

Russinovich, Mark, Manuel Costa, Cédric Fournet, David Chisnall, Antoine Delignat-Lavaud, Sylvan Clebsch, Kapil Vaswani, and Vikas Bhatia. 2021. Confidential Computing: Hardware-Based Trusted Execution for Applications and Data. *Communications of the ACM* 64 (6): 54–61. doi:https://doi.org/10.1145/3453930.

Serger, Sylvia Schwaag, and Igor Martins. 2023. An Age of Disentangled Research? *Issues in Science and Technology*, 40 (1): 38–43. https://doi.org/10.58875/wbcq3772

Shih, Tommy, Alex Chubb, and Diarmuid Cooney-O'Donoghue 2023. Scientific Collaboration Amid Geopolitical Tensions: A Comparison of Sweden and Australia. *Higher Education*, 87, 1339–1356 doi:https://doi.org/10.1007/s10734-023-01066-0.

Trillion Parameter Consortium. (2024). *Open platform for AI development.* https://www.anl.gov/trillion-parameter-consortium

University Research Compliance. (2024). *Technology control plans and export controls.* https://research-compliance.umich.edu/export-controls

U.S. Bureau of Industry and Security. (2024). *Commerce control list additions and revisions: Implementation of controls on advanced technologies.* https://www.federalregister.gov/documents/2024/09/06/2024-19633/commerce-control-list-additions-and-revisions-implementation-of-controls-on-advanced-technologies

Wagner, Caroline S., Koen Jonkers, Han Woo Park, and Loet Leydesdorff. 2015. The Continuing Growth of Global Cooperation Networks in Research: A Conundrum for National Governments. *PLoS ONE* 10 (7), e0131816. doi:https://doi.org/10.1371/journal.pone.0131816.

Yang, Qiang, Yang Liu, Tianjian Chen, and Yongxin Tong. 2019. Federated Machine Learning: Concept and Applications. *ACM Transactions on Intelligent Systems and Technology* 10 (2), 1–19. doi:https://doi.org/10.1145/3298981.

# Chapter 9
# Game-Changing Businesses

**Yousif Elsamani**

## 1 The Innovation Paradox

Innovation has become one of the most powerful words of our time. Governments, corporations, universities, and entrepreneurs increasingly portray innovation as the essential engine of progress, capable of solving humanity's most urgent challenges—from climate instability and healthcare inequities to economic stagnation and social fragmentation. Every major institution claims to be pursuing innovation, and every technological advance is celebrated as a breakthrough. Yet beneath this rhetoric lies a growing contradiction: despite record levels of research activity, investment, and technological advancement, the real transformative impact of innovation appears to be slowing rather than accelerating.

This paradox drove the discussion on "Game-Changing Businesses," during the 2025 Science and Technology in Society (STS) *forum*. The conversation examined a deceptively simple question: *why do so many technological breakthroughs fail to become game-changing businesses—and what must change for innovation to translate into real societal impact?*

Session chair André Kudelski articulated the central challenge with precision, reminding the audience that invention alone is never enough to reshape industries or society. As he stated,

> Just having something new will not create something fundamentally different... When you can find the right formula between a new technology, the right business model, and the scalability... then you have something that is extremely different.—**André Kudelski**

Empirical evidence increasingly supports this perspective. A landmark analysis by Park, Leahey, and Funk examined 45 million scientific papers and 3.9 million patents across six decades, applying a novel metric for breakthrough potential, the CD

Y. Elsamani (✉)
Institute for Future Initiatives, The University of Tokyo, Tokyo, Japan



H. Komiyama et al. (eds.), *AI, Innovation, and Beyond*,
https://doi.org/10.1007/978-981-95-7020-1_9

(Convention Disruption) index (Park et al., 2023). Their findings are striking: although the global research system produces exponentially more scientific and technological output every year, these outputs have become significantly less disruptive over time. Scientific papers and patents today are measurably less likely to redirect fields, overturn existing knowledge, or generate transformative new pathways. The decline is consistent across domains, from physics and chemistry to medicine, computing, and engineering, suggesting a structural shift rather than a field-specific slowdown. In short, we are generating more innovation-related activity, but less innovation-related impact.

A complementary body of research reveals a parallel dynamic in the economics of technological progress. Bloom, Jones, Van Reenen, and Webb show that modern research requires exponentially larger teams and investments to achieve even incremental improvements (Bloom et al., 2020). Their analysis of Moore's Law demonstrates that the number of researchers needed to maintain the historic pace of transistor density improvement is now over 18 times higher than in earlier decades. Across multiple industries, the same pattern emerges: ideas are becoming harder to find, and each additional unit of R&D produces a smaller marginal return. The assumption that larger investments in research automatically produce more innovation is increasingly misaligned with empirical reality. As efforts in innovation expands, transformative output paradoxically contracts.

This paradox raises an unsettling question: If technological development is accelerating in volume and complexity, why is it slowing in real-world impact? A growing consensus suggests that scientific and technological breakthroughs alone cannot deliver disruptive transformation. They must be integrated into the changes in business models, organizational capabilities, social systems, and value-creating ecosystems.

The mobile phone, a case referenced in the session, illustrates this dynamic clearly. As Kudelski noted, the early "Motorola" phones were technically impressive but "transportable" rather than truly mobile, and "heavy on your budget". Adoption was limited because the technology lacked the right pricing points to be deployed in mass. Only when size, cost, and network infrastructure aligned with consumer expectations did usage accelerate exponentially. Seen through the lens of Everett Rogers' seminal Diffusion of Innovations framework, adoption accelerates when the perceived relative advantage, compatibility, complexity, observability, and trialability align with user expectations (Rogers, 2003). Rogers argues that success depends less on the brilliance of the invention itself and more on the architecture of communication, usability, and timing. Innovation becomes transformative not when technology becomes possible, but when it becomes desirable and accessible at scale.

This tension also appears in Clayton Christensen's refinement of disruptive innovation theory (Christensen et al., 2015). Christensen, Raynor, and McDonald emphasize that disruption does not arise simply from superior technology; it occurs when new entrants restructure markets by targeting underserved or overlooked segments. Many organizations mislabel sustaining improvements by remarkable advancements of technologies as "disruptive," reinforcing incrementalism rather than transformation. This confusion contributes directly to the innovation paradox: we celebrate technical progress but fail to redesign the market or societal structures required to convert invention into transformation.

At the organizational level, James March provides a powerful explanation for why innovation systems drift toward incremental improvement rather than radical renewal. His classic distinction between exploration (experimentation and risk) and exploitation (refinement and efficiency) explains that organizations naturally prioritize exploitation because it yields predictable, measurable results (March, 1991). Over time, this preference for certainty produces systems that are optimized for short-term performance but structurally resistant to breakthrough innovation. March warns that organizations that rely overwhelmingly on exploitation may become "effective in the short run, but self-destructive in the long run". When viewed alongside the macro-level research of Park et al. (2023) and Bloom et al. (2020), it becomes clear that the innovation paradox is not a failure of creativity—it is a failure of system design.

This dynamic also aligns with broader cultural and strategic pressures described in the session. Around the world, companies invest heavily in R&D but struggle to transform those investments into new growth curves. Breakthrough inventions become trapped within old business models. Promising technologies are developed but fail to scale. The gap between technological possibility and societal adoption widens. The world is not short of innovation; it is short of mechanisms that make innovation transformative.

This leads to the defining question posed by Kudelski and echoed throughout the panel discussion: *How can innovation translate into game-changing businesses?*

The leaders on stage—from cinema and digital entertainment to cloud computing, platform business, and renewable energy—argued that innovation becomes world-changing only when it is integrated into viable business models, organizational redesign, ecosystem coordination, and societal values. Technology becomes meaningful when it changes incentives and relationships, when it expands rather than compresses opportunity, and when it allows collaboration to achieve "incredible things". This requires strategic choice, leadership courage, and a willingness to rethink how industries and institutions operate.

As Kudelski framed the challenge, the world needs new approaches to accelerate the translation of scientific knowledge into social and economic values. The rest of this chapter explores how industries that faced existential threats reinvented themselves and how organizational and societal systems must evolve to turn breakthrough inventions into breakthrough realities.

To understand what makes innovation truly transformative, we turn to real-world cases where industries repeatedly reinvented themselves, cinema, digital platforms, creative technologies, and artificial intelligence, and examine how leaders converted technological possibilities into enduring business and societal impact.

## 2 Reinventing Industries: Lessons from Cinema, Digital Platforms, and AI

The Innovation Paradox explains why many technological breakthroughs do not become real transformations. The next question is what distinguishes industries and organizations that repeatedly reinvent themselves from those that fade. The discussion highlighted a powerful insight: transformative innovation rarely depends on technology alone. Instead, it emerges when organizations and industries reshape business models, creative cultures, and ecosystem structures to unlock new forms of value. Whether in cinema, digital entertainment, cloud platforms, or artificial intelligence, the pattern is consistent—game-changing businesses redesign the rules of participation, enabling new value chains and redistributing capability, access, and creativity.

The cinema and digital entertainment industries provide a vivid example of this capacity for reinvention. As Elizabeth Daley, Dean of the USC School of Cinematic Arts, observed, the film industry has endured a century of disruption, from sound and color to television, home video, digital production, streaming, and AI. Every wave of technological change brought predictions of collapse.

> Every time a new technology has come along—sound, television, streaming, digital capture—we were told it was the demise of cinema. None of that happened.—**Elizabeth Daley**

What kept the film industry alive was not defensive resistance but its capacity to absorb new technologies into its mission: storytelling. Daley emphasized that the value of technology in cinema is judged not by novelty, but by whether it strengthens creative impact.

> People pay us for making impactful stories, and as long as technology enables us to do that, we will adopt it. If it doesn't, we will abandon it immediately.—**Elizabeth Daley**

This perspective echoes Teece's argument that technological superiority is insufficient unless accompanied by business model innovation capable of capturing and delivering value (Teece, 2010). As Teece explains, sustainable advantage arises not from invention alone but from the design of the value creation, delivery, and capture architecture that surrounds it. The film industry's repeated reinvention demonstrates what Demil and Lecocq describe as dynamic consistency, in which business models continually adjust their internal components in response to environmental shifts (Demil & Lecocq, 2010). Rather than treating each technological wave as a threat, cinema repeatedly reconfigures its practices—production, distribution, financing, intellectual property protection—to remain aligned with its purpose. Reinvention becomes a capability, not an event.

A similar pattern emerges in the transformation of Sony from a hardware-centric company to a global entertainment and creative technology platform. Hiroaki Kitano explained that

> Fifteen years ago, almost 90% of our revenue was in hardware. Today, 75% comes from entertainment: games, music, movies, and sports. It is a dramatic transformation.—**Hiroaki Kitano**

Sony's reinvention was not accomplished by technology alone, but by redefining identity and purpose.

> We are here for creators. That became our true north.—**Hiroaki Kitano**

This represents a reversal of the traditional innovation sequence: rather than pushing new hardware into the market and hoping for adoption, Sony defined its desired outcome—empowering creators—and reconfigured its technological capabilities to serve that mission.

Behind this transformation lies the capability to absorb, integrate, and recombine diverse domains of knowledge—what Cohen and Levinthal describe as absorptive capacity (Cohen & Levinthal, 1990). Their research shows that organizations must accumulate and internalize the right knowledge stocks in order to recognize the value of external inputs and translate them into commercial innovation. Sony's deep roots in hardware engineering allowed it to understand and support creator needs in software, content, and interactive media—bridging domains that competitors treated separately. This is consistent with Hanelt et al.'s observation that digital transformation requires holistic co-evolution across technologies, business processes, and organizational culture, rather than isolated digital implementation (Hanelt et al., 2021). Sony's evolution embodies that principle: it built an ecosystem where creators, technology, distribution, and audience experience are tightly integrated.

Yet perhaps the clearest example of redefining an industry through business model reinvention is Amazon, whose journey illustrates the profound difference between product innovation and platform innovation. As Kim Majerus, Vice President at Amazon Web Services (AWS), reflected,

> Over 30 years ago, Jeff Bezos founded one of the most game-changing businesses in history when he created Amazon.com.—**Kim Majerus**

What began as an online bookstore evolved into a logistics powerhouse and ultimately into AWS, the cloud computing platform that enabled thousands of new companies to emerge. Rather than optimizing a retail business, Amazon reimagined the architecture of digital infrastructure itself. AWS was not merely a technological breakthrough—it was a radical new business model in computing, converting fixed-cost data centers into elastic pay-as-you-go utility services. This aligns closely with Casadesus-Masanell and Zhu's argument that business model innovation can generate more value than the technology itself, creating strategic dilemmas for incumbents who struggle to compete with new models built around different economic logics (Casadesus-Masanell & Zhu, 2013).

Majerus emphasized that Amazon's guiding logic is grounded in application rather than technology itself:

> AI for AI's sake is not the key to innovation. It's really how we are going to use AI as a tool across the domains... that will actually define and decide the impact.—**Kim Majerus**

Her message reinforces a recurring theme across the panel: innovation becomes transformative only when it expands opportunity and capability for others. Amazon's platform model exemplifies what we can call digital leadership competencies, transforming business models, processes, and culture simultaneously rather than layering technology on existing structures.

The same reckoning is now unfolding with artificial intelligence, especially generative AI. The future of creative industries may be shaped not only by AI's ability to enhance individual creativity but also by its potential to narrow collective diversity, as demonstrated by Doshi and Hauser, who found that while AI-assisted writing improves assessed creativity, outputs converge around similar patterns (Doshi & Hauser, 2024). This raises profound questions about authorship, originality, and the cultural role of creativity—issues Daley and Kitano both highlighted, signaling that the next wave of reinvention will depend not only on technology or business models, but values, governance, and purpose.

Together, these case studies reveal a powerful pattern. Whether in cinema, digital entertainment, platform computing, or AI, game-changing businesses succeed by reimagining value chains, realigning organizational purpose, and designing ecosystems where technology enables human possibility rather than replacing it. Innovation becomes transformative when leaders choose reinvention over protectionism, flexibility over rigidity, and collaboration over isolation. As Daley reminded the audience,

> Do not put people, companies, and technologies in silos. Let us move forward together.—**Elizabeth Daley**

The next frontier where this transformation logic is most urgently tested is energy, where system-level redesign—not incremental improvement—will determine whether innovation can reshape the material foundations of society.

## 3 Energy Transition as System Transformation

Digital platforms, cinema, and creative technologies show how industries can reinvent themselves by redesigning business models and ecosystems. The energy sector represents a far greater challenge—one where innovation demands system-level transformation rather than product-level improvement. While digital and creative industries operate in rapidly evolving market environments, the energy sector must navigate technological complexity, regulatory structures, physical infrastructure limitations, and geopolitical constraints. The shift from centralized fossil fuel systems to distributed, renewable energy sources is not simply a technological transition; it requires re-architecting the way modern societies function. In this domain, innovation becomes transformative only when it aligns technical breakthroughs with new mechanisms for coordination, market design, and value creation.

The session's discussion revealed how profoundly energy systems have changed within a short period. As Yusuke Toyoda, CEO of Digital Grid, explained, Japan alone has experienced an explosion of decentralized renewable installations in recent years, dramatically altering the operational dynamics of its electrical grids. It introduces a new challenge: renewable energy sources fluctuate from minute to minute based on weather and local conditions, requiring constant recalibration across the system. Toyoda emphasized the urgency of this complexity, and necessity of platform technology changing energy system and business.

> Renewables fluctuate every minute, every second. What we need is platform technologies allowing flexibility of distributed renewable energy systems.—**Yusuke Toyoda**

Unlike traditional power generation, where output is stable and large units can be coordinated centrally, renewable systems require network-level responsiveness, real-time data integration, and new patterns of exchange among hundreds of thousands of distributed producers and consumers. The problem is no longer simply how to generate energy, but how to orchestrate it, which fundamentally transforms the nature of innovation in the sector.

This shift illustrates the critical insight of Rantala et al., who demonstrate that innovation adoption in complex systems is driven not only by technological performance but by the alignment of economic, environmental, and institutional value dimensions (Rantala et al., 2018). Energy innovation succeeds when it improves efficiency and sustainability simultaneously and when it integrates institutional actors such as regulators, market operators, equipment manufacturers, and local communities. When values are aligned, innovation accelerates; when misaligned, progress stalls. The energy transition further validates the need for a bidirectional approach. It is not driven simply by the invention of solar panels, but by a societal mandate for sustainability that forces the redesign of market institutions and dictates the specific "platform technologies" required to stabilize the grid.

Energy embodies this principle: adding solar panels or batteries is not transformative in isolation—what is transformative is the redesign of market structure, pricing mechanisms, and control systems that allow distributed resources to act as coordinated networks rather than disconnected assets. The energy transition, therefore, is as much an exercise in organizational and institutional innovation as it is in technological advancement.

This perspective aligns with Westerman, Bonnet, and McAfee's conclusion that digital leadership requires the simultaneous transformation of business models, operational processes, and organizational culture (Westerman et al., 2014). Digital Grid exemplifies this approach. It is not simply deploying new technologies but enabling a new marketplace where energy becomes an actively traded digital asset rather than a passive physical commodity. The company's innovation reframes the roles of producers and consumers, transforming households, businesses, and communities into active participants in energy exchange. This shift creates a fundamentally different logic of coordination—what Toyoda and his colleagues describe as the democratization of the electricity economy.

Seen through the lens of the broader conference theme, the energy sector reveals a core principle of game-changing business: true transformation occurs when innovation reconfigures systems rather than components. Just as cinema reinvented storytelling and distribution, and Amazon reinvented infrastructure and access, the energy transition requires new architectures of interaction and integration. Technologies such as solar, wind, or energy storage are meaningful only when embedded in systems capable of coordinating complex, decentralized activity. The value emerges not from the device, but from the platform that connects them.

The session's energy discussion also highlights the societal stakes of innovation. Unlike digital transformation or creative technology disruption, energy fundamentally shapes the capacity of communities, industries, and nations to function and grow. Energy innovation must therefore manage risk, reliability, and public trust at levels unmatched in most other sectors. It is here that the Innovation Paradox becomes most visible: technological capability already exists, but adoption lags because governance, markets, and coordination mechanisms have not yet caught up.

For energy, as for every sector represented in the discussion, technology becomes transformative only when the surrounding system evolves with it—when business models, policies, skills, and societal values align. The energy transition stands as both a warning and an aspiration: a reminder that complex innovations cannot succeed through invention alone, and proof that game-changing business requires systemic thinking, cross-sector collaboration, and new ways of organizing value.

Just as energy transformation requires rethinking system coordination and institutional architecture, the rise of AI requires a comparable redesign—not of physical infrastructure, but of human capability and organizational structure. The same system-level transformation logic applies to the future of work and talent in an AI-driven economy, where the stakes are not infrastructure reliability but human capability, opportunity, and inclusion. As artificial intelligence reshapes industries, the question is no longer whether jobs will disappear, but how organizations, workers, and societies will adapt to thrive. In all these domains, innovation quality will increasingly be judged not only by its economic returns, but by its contribution to resilience, inclusion, and societal well-being.

## 4 The Future of Work and Talent in an AI World

While energy transition reshapes the physical foundations of society, an equally profound transformation is underway in the world of work. Artificial intelligence is reshaping industries not only through automation but through the redefinition of tasks, skills, and human roles. While public debate often frames AI as a threat to employment, the panel discussion emphasized a far more nuanced reality: the future of work will be defined not by the displacement of human labor, but by the recomposition of work around new forms of collaboration between human capability and machine intelligence.

The central challenge is not the elimination of jobs, but the readiness of individuals, organizations, and societies to adapt. This perspective echoes the work of David Autor, who argues that the historical narrative of automation has been repeatedly misinterpreted (Autor, 2015). Autor demonstrates that technological change does not simply replace workers; it also complements human labor, enabling higher productivity and expanding economic activity in ways that create new jobs and redesign existing ones (Autor, 2015). Automation substitutes for routine, codifiable tasks while amplifying human strengths in creativity, judgment, adaptability, and collaboration. What changes is the content of work, not the fundamental need for human participation. Harrick Vin articulated this shift as a move from efficiency to expertise:

> We are looking at... activity centric AI, using AI to improve productivity of tasks. Tomorrow, we'll have to think about human centric or organization centric AI, which will offer wisdom and expertise at your fingertips. It will improve your ability to judge and make decisions.—**Harrick Vin**

Autor notes that anxieties about mass technological unemployment have appeared in every major wave of invention and have consistently proven inaccurate (Autor, 2015). Yet his research also highlights a crucial caveat: while total employment may not decline, work will shift dramatically, and workers must be prepared to evolve with it.

The panel discussion underscored this essential truth. Kim Majerus emphasized that the real challenge is not whether AI will be adopted, but who will be ready to use it. The statement captures the heart of the future-work transformation: value will not be created by developing AI technologies in isolation, but by integrating them effectively into workflows that enhance human potential. Majerus stressed that the workforce of the future requires new forms of capability-building that go beyond technical expertise, remarking,

> We must focus on developing not just technical proficiency but also embracing the unique human traits which AI can never replace: creativity, emotional intelligence, and problem solving.—**Kim Majerus**

Harrick Vin, Chief Technology Officer at Tata Consultancy Services (TCS), reinforced this view, rejecting the binary narrative of replacement in favor of augmentation.

> We are rapidly getting into an era of a hybrid workforce, where people and machines will work shoulder to shoulder... This virtuous cycle of mutual augmentation is going to double the value generation or work capacity.—**Harrick Vin**

These insights resonate with Cohen and Levinthal's concept of absorptive capacity, which defines an organization's ability to recognize, assimilate, and apply new knowledge as a function of its existing knowledge base (Cohen & Levinthal, 1990). Their framework suggests that simply providing access to AI tools is insufficient; organizations and individuals must first develop the cognitive foundations required to understand and meaningfully integrate new technologies. Without absorptive capacity, AI adoption becomes superficial and performative rather than transformative.

Thus, learning becomes not a process of training workers to follow instructions, but of equipping them to interpret, adapt, and innovate. Developing this capacity requires organizations to navigate the tension between exploitation—optimizing existing operations—and exploration—experimenting with new technologies and ways of working. This challenge aligns with the concept of organizational ambidexterity, which Sebastian Raisch et.al identify as the ability to pursue incremental and radical innovation simultaneously (Raisch et al., 2009). Their research demonstrates that ambidextrous organizations significantly outperform those that commit exclusively to efficiency or exclusively to experimentation. Firms that succeed in the age of AI will be those that cultivate space for exploration while sustaining operational excellence.

Birkinshaw further shows that ambidexterity requires deliberate structural choices—such as separate units for exploratory work—as well as mechanisms for knowledge transfer and leadership alignment (Birkinshaw & Gibson, 2004). These insights echo the panel's recognition that AI adoption requires both organizational redesign and cultural evolution, not merely investment in new tools. In a world where AI accelerates change, the most valuable organizational capability becomes the ability to learn faster than the environment is changing, and to convert that learning into coordinated action.

Yet the benefits of AI also carry risks. If technologies enable individual workers to become more productive while simultaneously reducing diversity of ideas and outputs, organizations may face a new form of creative narrowing. Research by Doshi and Hauser reveals that while generative AI improves subjective evaluations of creative writing, it also produces more homogeneous and less novel outputs (Doshi & Hauser, 2024). This creates a paradox in which AI empowers individuals but may diminish the originality of collective output.

This finding parallel concerns within creative industries raised in the panel: AI might elevate average performance while compressing exceptional and distinctive contributions. Innovation could become easier to achieve, but harder to differentiate. This risk extends beyond artistic creation to all fields where originality, strategic decision-making, and human perspective matter. If creativity converges and decision-making becomes standardized, organizations may sacrifice precisely the kinds of judgment and distinctiveness that competitive advantage requires.

The challenge, then, is not only to enable AI-enhanced productivity, but to protect and cultivate human uniqueness. The panelists agreed that the future of work must be shaped intentionally rather than reactively. As AI becomes embedded in more tools, platforms, and workflows, societies face a crucial choice: whether to let automation deepen inequality or to design it to expand opportunity. The transition demands leadership capable of building cultures of learning, redesigning work around human strengths, and ensuring that technological acceleration translates into shared progress.

The stakes are as societal as they are organizational: the future of work will determine the distribution of dignity, purpose, and participation in the decades ahead. If AI forces companies to rethink how work functions and how capability is developed, then the next challenge is equally urgent: how leaders confront uncertainty, risk, and responsibility while pursuing innovation at scale. The conversation in the session turned toward what keeps leaders awake and what dreams motivate them forward.

## 5 Risks, Responsibilities, and Dreams for the Future

As the panel turned from examining industry transformations to reflecting on the future, a common theme emerged: the most difficult challenges of innovation are not technological, but human, organizational, and societal factors. Innovation requires leaders to confront uncertainty, make decisions without full control, and balance the tension between accelerating transformation and protecting stability. The discussion revealed that what keeps leaders awake is not the fear of new technologies, but the responsibility to ensure that innovation creates more value than it destroys—and that the benefits are shared broadly rather than concentrated narrowly.

This tension reflects the deeper structural challenge of navigating what James March earlier described as the fundamental dilemma between exploration and exploitation (March, 1991). Leaders must invest in experimentation, risk-taking, and long-term exploration, while simultaneously maintaining operational performance and stakeholder confidence. Organizations that lean too heavily toward exploitation may excel temporarily but stagnate over time; those that focus exclusively on exploration may collapse under uncertainty. In rapidly shifting technological environments, the responsibility of leadership is to manage this tension intentionally, designing organizations that can evolve continuously without losing coherence.

This need for balance lies at the heart of organizational ambidexterity, which O'Reilly and Tushman identify as a defining characteristic of successful organizations in environments of rapid change (O'Reilly & Tushman, 2013, 2011). Their research demonstrates that ambidextrous organizations achieve significantly better performance because they create structures that separate exploratory and exploitative activities while maintaining senior leadership oversight to align them strategically. Ambidexterity, in this sense, becomes a dynamic capability—the ability to reconfigure resources, processes, and strategies repeatedly over time (O'Reilly & Tushman, 2008). It allows organizations to pursue innovation aggressively while sustaining the operational excellence necessary to compete in the present.

The panelists articulated this leadership challenge with clarity and urgency. André Kudelski framed the issue as a societal responsibility, emphasizing that innovation must be directed toward positive impact rather than disruption for its own sake:

> How do we accelerate technology adoption in a way that positively changes the world?—
> **André Kudelski**

His question reflects a central ethical consideration: innovation must be evaluated not only on technical feasibility and commercial potential, but on its ability to generate net value—economic, social, and human. This perspective aligns with Westerman, Bonnet, and McAfee's argument that successful digital transformation requires cultural and governance transformation alongside technological investment (Westerman et al., 2014). Technology is the catalyst, but leadership defines the outcome.

From a different vantage point, Hiroaki Kitano described a future where AI dramatically changes the structure of capitalism and the scale of organizations, raising both opportunity and concern:

> We will see companies selling billions and run by five or ten people. This will completely change the way capitalism works.—**Hiroaki Kitano**

This prediction underscores a profound risk: that technological acceleration could concentrate power and opportunity within increasingly small groups, widening divides between those who control technology and those excluded from its benefits. The potential for exponential productivity gains is extraordinary, yet so is the potential for societal imbalance. Innovation becomes a question of governance, not just capability.

Elizabeth Daley echoed this theme, emphasizing that fragmentation and isolation are barriers to progress across industries and nations. She warned against treating innovation as the domain of isolated experts or siloed industries. Her insight reinforces the idea that collaborative ecosystems, rather than competitive silos, will define the next wave of transformative innovation. The challenges ahead—whether energy transition, AI ethics, or economic restructuring—are not solvable within traditional boundaries. They demand cross-sector alliances and new institutional architectures.

Yet despite the risks and uncertainties described, the conversation was ultimately optimistic. The panelists agreed that innovation remains the most powerful force for progress when guided by human values, ethical responsibility, and shared purpose. Leadership, in this context, is not merely strategic but aspirational—the ability to articulate a vision worth pursuing, to protect possibility in the face of risk, and to inspire collaboration despite uncertainty.

Innovation becomes game-changing not when it is technologically radical, but when leaders commit to dreams large enough to matter and disciplined enough to succeed. Technology reshapes industries; leadership reshapes society.

The final question, then, is not whether technology will advance, but whether we will build the systems, models, and values required to turn innovation into real impact. The concluding section reflects on the lessons emerging from the session and synthesizes the conditions necessary for transforming invention into societal progress.

## 6 Conclusion: From Invention to Impact

Across the diverse perspectives shared in the session "Game-Changing Businesses," a unifying message becomes clear: the world is not suffering from a shortage of innovation, but from a shortage of transformation. Breakthrough discoveries, unprecedented computing power, accelerating research output, and vast technological potential surround us. Yet as the evidence shows, the ability of innovation to fundamentally reshape industries, economies, and society depends not on the brilliance of inventions alone, but on the systems that enable them to scale.

Transformation is not achieved through technological novelty, but through alignment—across business models, organizational capabilities, markets, and societal values. The Innovation Paradox introduced in the opening section reveals that while the production of scientific knowledge has grown exponentially, the disruptive impact of innovation is declining. Organizations and societies generate more ideas than ever, yet struggle to translate them into breakthroughs that shift value, redesign markets, or solve urgent challenges.

This is not a failure of creativity; it is a failure of alignment—between technology and business models, invention and adoption, exploration and exploitation, and individual capability and institutional readiness.

The industry cases explored throughout this chapter illustrate that transformation succeeds when innovation is accompanied by reinvention. Cinema survived a century of disruption because it integrated new technologies into a deeper purpose: storytelling. Sony redefined itself from a hardware company to a creative platform by realigning identity around creators and building absorptive capacity. Amazon reimagined digital infrastructure, enabling others to innovate on its platform, demonstrating the exponential leverage of business model innovation. The energy transition revealed that innovation affecting the physical foundations of society requires full system redesign. And the discussion on AI and the future of work underscored that technology alone cannot deliver progress without structures to build capability, inclusion, and adaptability.

Across these domains, one lesson stands out: game-changing businesses succeed when technology, business models, and broader societal systems evolve in coherence. Breakthrough impact emerges when organizations build business models capable of capturing and delivering new forms of value; cultivate cultures that balance optimization with exploration; develop talent and organizational capabilities that convert knowledge into creativity; and intentionally design ecosystems where incentives, policies, and governance support adoption and scale. This principle is visually summarized in Fig. 9.1, which illustrates the bidirectional nature of game-changing business. While traditional views treat innovation as a linear flow from discovery to outcome, game-changing transformations often proceed in reverse: they start by defining the desired societal and economic outcomes (such as decarbonization or creative empowerment) and backcast to design the necessary ecosystems, business models, and scientific targets. True impact requires a continuous loop where societal needs guide scientific discovery just as powerfully as discovery reshapes society.

As André Kudelski noted, while technology allows us to do "incredible things," we must be vigilant to ensure it does not create "major issue" or "abomination". Technology is neutral; impact is intentional. The risks ahead are real.

The future could bring shared prosperity and democratized creativity, or concentrated power and deepened inequality. Elizabeth Daley underscored that collaboration, rather than fragmentation, must guide the path forward. Ultimately, innovation must be evaluated not by novelty, but by consequence. Not by how fast we move, but by what we move toward. Transformation is possible only when leaders build ecosystems capable of turning invention into impact, and when ambition is matched by responsibility.

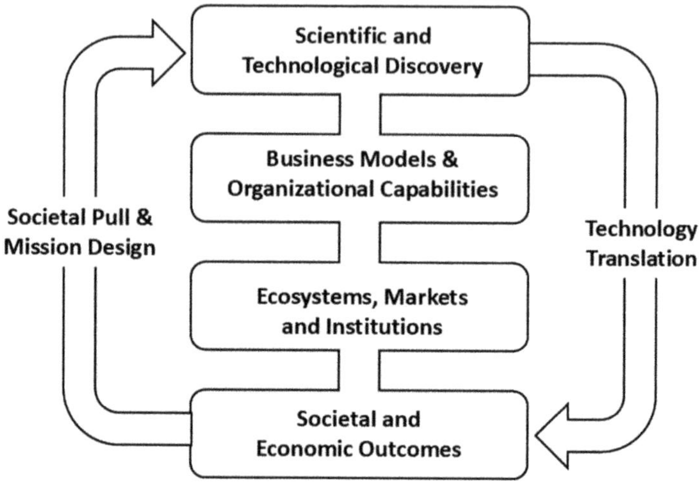


**Fig. 9.1** The bidirectional cycle of game-changing business. Innovation is not merely a linear translation of science into society. It is a continuous feedback loop where societal and economic goals (such as sustainability or inclusion) actively shape business model design and direct scientific discovery. Game-changing businesses align these forces, ensuring that technological push meets societal pull

The chapter closes with the insight that anchored the session's conclusion, a reminder that in a rapidly evolving world, the greatest risk is standing still:

> We should change if we do not want to be changed or replaced.—**André Kudelski**

The task ahead is to build the systems, cultures, and collaborations that allow innovation to fulfill that promise, aligning invention, commercialization, and societal purpose. The future belongs to those who can bridge imagination with execution, risk with responsibility, and vision with system design, turning invention into meaningful impact.

**Competing Interests** No Competing interests.

## Session Panelists

**André Kudelski** is Chairman and CEO of Kudelski Group. He is also President of Innosuisse, the Swiss Federal Agency for Innovation, and Vice-Chairman of the SwissAmerican Chamber of Commerce. He acts as Lead Director to the Board of Directors of the Publicis Groupe and is a member of the Strategic Advisory Board of École Polytechnique Fédérale de Lausanne (EPFL). He is a Council member of the Science and Technology in Society (STS) *forum*, Japan. He previously served on the Boards of Nestlé, HSBC Private Banking Holdings (Suisse) and Dassault Systèmes. André Kudelski has a MSc in Applied Physics from EPFL and is a member of the Swiss Academy of Engineering Sciences (SATW).

**Elizabeth M. Daley** has served as Dean of the USC School of Cinematic Arts for 34 years. Under her leadership, the school has added three new divisions; established the Entertainment Technology Center, an industry think tank; built the Robert Zemeckis Center for Digital Arts; and established 36 endowed positions. She serves on the Board of Directors of Digital Domain and the Benton Foundation and is a former chair of the World Economic Forum's Global Agenda Council on Media, Entertainment and Information. She is a member of the Directors Guild of America and the Academy of Motion Picture Arts and Sciences. She holds a Ph.D. from the University of Wisconsin, M.A. and B.A. degrees from Tulane University, and an honorary Doctorate of Letters from Hong Kong Baptist University.

**Hiroaki Kitano** is Chief Technology Fellow at Sony Group Corporation. He served as Chief Technology Officer of the company from 2022 to 2025. He joined Sony Computer Science Laboratories (Sony CSL) in 1993, serving as President and CEO of the company since 2011. In 2024, Kitano established the Sony Women in Technology Award with Nature as one of the Principal Founders. Outside Sony, Kitano is a member of the OECD Expert Group on AI Futures, Japan's AI Strategy Council and AI Safety Institute. Within academia, he serves as a professor at Okinawa Institute of Science and Technology (OIST). He is the Founding President of RoboCup Federation. In 2021, Kitano established the Nobel Turing Challenge, a grand challenge to develop a new engine for scientific discovery.

**Kim Majerus** leads the global education and U.S. state & local government vertical at Amazon Web Services (AWS). Her sales organization includes EdTech and GovTech independent software vendors (ISVs), solution and vertical strategy leaders who focus on Health and Human Services, Justice and Public Safety, Transportation and the Internet of Things for state and local government customers. Prior to joining AWS, Kim spent 16 years at Cisco Systems, most recently as Vice President of Cisco Public Sector for the U.S. State, Local and Education Market Segment. Prior to moving to public sector, Kim held positions in engineering, service provider sales, partner channels, global channels and as General Manager for Ireland. Kim dedicates much of her time to mentoring young women and promoting STEAM education in her local community. She has received many honors, including Chicago Business Journal Women of Influence 2022, State Scoop's 2021 and 2022 Top 50 Industry Leader award, State Top 50 Women in Technology, YWCA Silicon Valley leaders, and Top 50 Business Leaders in Ireland. Kim, her husband Michael, and their two daughters live in Chicago, IL.

**Yusuke Toyoda** is the President and CEO of Digital Grid Co., Ltd., a role he has held since July 2019. He joined Digital Grid in 2018 as a founding member and contributed to launching the business. Earlier, he worked at Integral, a private equity fund, where he led investments across various sectors from 2016. Before that, he was at Goldman Sachs in the Securities Division, focusing on structuring and sales of foreign exchange and credit-related financial products, and later in the Strategic Investment Department, where he developed and invested in large-scale solar power projects. He holds a master's degree in Technology Management for Innovation from the University of Tokyo Graduate School of Engineering.

**Dr. Harrick Vin** is the SVP and Chief Technology Officer (CTO) at Tata Consultancy Services (TCS), where he leads the company's research and innovation strategy, focusing on future of technology and future of work. Dr Vin is a TCS Fellow, a Fellow of Indian National Academy of Engineering, and a Distinguished Alumnus of IIT Bombay. In 2025, he was recognized by Forbes India as one of the Top 30 Indians leading the AI revolution, and by MIT Sloan India as one of the Top 100 most influential people in tech and innovation. Prior to his current role, he has held several key leadership roles at TCS, including Chief Services Innovation Officer and Chief Scientist at Tata Research Development and Design Centre (TRDDC). He was the founder of Digitate, a TCS venture, and led the development of ignio, the world's first cognitive automation product for enterprise IT. Prior to joining TCS, he was a Professor of Computer Sciences at the University of Texas at Austin (USA) for 15 years. Harrick received his PhD in Computer Science and Engineering from the University of California, San Diego, and B.Tech. in Computer Science and Engineering from the Indian Institute of Technology (IIT), Bombay.

## References


Autor, D. H. (2015). Why Are There Still So Many Jobs? The History and Future of Workplace Automation. *Journal of Economic Perspectives, 29*(3), 3–30. https://doi.org/10.1257/jep.29.3.3

Birkinshaw, J., & Gibson, C. (2004). Building Ambidexterity Into an Organization. *MIT Sloan Management Review*. https://sloanreview.mit.edu/article/building-ambidexterity-into-an-organization/

Bloom, N., Jones, C. I., Van Reenen, J., & Webb, M. (2020). Are Ideas Getting Harder to Find? *American Economic Review, 110*(4), 1104–1144. https://doi.org/10.1257/aer.20180338

Casadesus-Masanell, R., & Zhu, F. (2013). Business model innovation and competitive imitation: The case of sponsor-based business models. *Strategic Management Journal, 34*(4), 464–482. https://doi.org/10.1002/smj.2022

Christensen, C. M., Raynor, M. E., & McDonald, R. (2015). What Is Disruptive Innovation? *Harvard Business Review*. Retrieved December 4, 2025, from https://hbr.org/2015/12/what-is-disruptive-innovation

Cohen, W. M., & Levinthal, D. A. (1990). Absorptive Capacity: A New Perspective on Learning and Innovation. *Administrative Science Quarterly, 35*(1), 128–152. https://doi.org/10.2307/2393553

Demil, B., & Lecocq, X. (2010). Business Model Evolution: In Search of Dynamic Consistency. *Long Range Planning, Business Models, 43*(2), 227–246. https://doi.org/10.1016/j.lrp.2010.02.004

Doshi, A. R., & Hauser, O. P. (2024). Generative AI enhances individual creativity but reduces the collective diversity of novel content. *Science Advances, 10*(28), eadn5290. https://doi.org/10.1126/sciadv.adn5290

Hanelt, A., Bohnsack, R., Marz, D., & Antunes Marante, C. (2021). A Systematic Review of the Literature on Digital Transformation: Insights and Implications for Strategy and Organizational Change. *Journal of Management Studies, 58*(5), 1159–1197. https://doi.org/10.1111/joms.12639

March, J. G. (1991). Exploration and Exploitation in Organizational Learning. *Organization Science, 2*(1), 71–87. https://doi.org/10.1287/orsc.2.1.71

O'Reilly, C. A., & Tushman, M. L. (2008). Ambidexterity as a dynamic capability: Resolving the innovator's dilemma. *Research in Organizational Behavior, 28*, 185–206. https://doi.org/10.1016/j.riob.2008.06.002

O'Reilly, C. A., & Tushman, M. L. (2013). Organizational Ambidexterity: Past, Present, and Future. *Academy of Management Perspectives, 27*(4), 324–338. https://doi.org/10.5465/amp.2013.0025

O'ReillyIII, C. A., & Tushman, M. L. (2011). Organizational Ambidexterity in Action: How Managers Explore and Exploit. *California Management Review, 53*(4), 5–22. https://doi.org/10.1525/cmr.2011.53.4.5

Park, M., Leahey, E., & Funk, R. J. (2023). Papers and patents are becoming less disruptive over time. *Nature, 613*(7942), 138–144. https://doi.org/10.1038/s41586-022-05543-x

Raisch, S., Birkinshaw, J., Probst, G., & Tushman, M. L. (2009). Organizational Ambidexterity: Balancing Exploitation and Exploration for Sustained Performance. *Organization Science, 20*(4), 685–695. https://doi.org/10.1287/orsc.1090.0428

Rantala, T., Ukko, J., Saunila, M., & Havukainen, J. (2018). The effect of sustainability in the adoption of technological, service, and business model innovations. *Journal of Cleaner Production, 172*, 46–55. https://doi.org/10.1016/j.jclepro.2017.10.009

Rogers, E. M. (2003). *Diffusion of Innovations, 5th Edition*. Simon and Schuster.

Teece, D. J. (2010). Business Models, Business Strategy and Innovation. *Long Range Planning, Business Models, 43*(2), 172–194. https://doi.org/10.1016/j.lrp.2009.07.003

Westerman, G., Bonnet, D., & McAfee, A. (2014). *Leading Digital: Turning Technology Into Business Transformation*. Harvard Business Press.

# Chapter 10
# Science Communication

**Cristian Mejia**

## 1 Introduction

We live in an era of unprecedented scientific achievement. From mRNA vaccines that saved millions of lives to artificial intelligence systems that rival human cognition, from gravitational wave detections that reveal the early universe to CRISPR technologies that rewrite the code of life itself—the pace and scope of scientific discovery have never been more impressive. Yet paradoxically, this golden age of scientific progress unfolds against a backdrop of deepening public skepticism and institutional fragility. As Mikihito Tanaka, panelist of the science communication session at the STS *forum*, warns, "three 'isms' are currently distorting science communication: populism, alarmism, and accelerationism."

The empirical evidence reveals a troubling landscape. While 76% of Americans report trust in scientists according to recent surveys (Pew Research Center, 2023), this figure represents not recovery but arrested decline—an 11-point drop occurred between 2020 and 2023 alone. More concerning still is the calcification of partisan divides: a 22-point gap now separates Democrats and Republicans in their trust of the scientific community (Funk & Kennedy, 2024). These numbers speak to something more profound than mere fluctuations in public opinion. They signal a fundamental crisis in the relationship between science and society, one with implications that extend far beyond the laboratory.

The stakes are highest precisely because this crisis is not merely about science. When trust in evidence-based reasoning erodes, both institutions—science and democracy—suffer in tandem. The principles that underpin rigorous scientific inquiry—respect for reliable evidence, commitment to rational argument, openness to courteous debate, and above all, dedication to truth—are the same principles that

C. Mejia (✉)
Institute for Future Initiatives, The University of Tokyo, Tokyo, Japan



H. Komiyama et al. (eds.), *AI, Innovation, and Beyond*,
https://doi.org/10.1007/978-981-95-7020-1_10

sustain healthy democratic deliberation. When these shared foundations crumble, the consequences cascade across society.

> The pursuit of truth is core to both a good democracy as well as to science… We are seeing trends in the world where little or no respect is being given to those sorts of arguments—little respect for reliable evidence, for rational argument, and certainly not for courteous debate.—**Paul Nurse**

The current moment offers stark illustration. In the United States, political leadership has increasingly demonstrated this disregard. This manifests not as benign neglect but as active suppression: "We're seeing in the US a silencing of the ability to obtain evidence about areas that the political leadership does not want to see being pursued," as Nurse observes. When political power seeks to determine which questions science may ask and which findings it may report, the threat ex-tends beyond any single nation's borders. What emerges in one democracy can metastasize in others, making the challenge of effective science communication not merely urgent but existential.

For decades, science communication operated under a seductive assumption: public skepticism toward science stems primarily from ignorance. According to this "deficit model," the solution was straightforward—provide more information, craft clearer explanations, improve science education. If people simply knew what scientists know, the reasoning went, they would naturally align their views with scientific consensus. The problem was framed as a knowledge deficit, and the remedy as knowledge transfer.

Recent evidence has delivered a decisive verdict: this model fails. Consider the most rigorous test to date. Gligorić and colleagues recruited 7800 participants and deployed five theory-based interventions specifically designed to increase conservative trust in scientists. The interventions drew on established psychological theories and targeted known mechanisms of distrust. The result? Zero measurable effect across all five approaches. Not modest effects, not promising trends—complete failure to move the needle on trust.

The deficit model's inadequacy becomes even more apparent when we examine what researchers call the "smart idiot effect." Among Republicans with higher science literacy, acceptance of climate consensus actually decreases rather than increases (Kahan et al., 2012). Greater knowledge, in other words, does not guarantee agreement—it can amplify disagreement when identity and values are at stake. This pattern reveals the true culprit: motivated reasoning. People process scientific information not to optimize accuracy but to protect their social identity and rein-force their group membership (Kahan, 2017). Facts alone cannot penetrate defenses erected to maintain communal bonds.

The depth of the trust deficit becomes clear when we examine public perceptions of scientists themselves. According to recent surveys, 43% of Americans believe that scientists adjust their findings to meet predetermined conclusions (Wellcome Trust, 2024). Nearly half the population, in other words, suspects researchers of cherry-picking results. Even more troubling, only 53% believe scientists are even capable of unbiased analysis (National Academy of Sciences, 2024). When substantial portions of the public question not just the conclusions but the very capacity for objectivity, no amount of additional information will bridge the divide.

Yet remarkably, the deficit model persists. Tanaka identifies this as the danger of "alarmism"—when communicators encounter misinformation and disinformation, they reflexively respond with intensified efforts toward enlightenment of "correct" scientific knowledge. This response, however well-intentioned, reinforces the deficit model—a model we thought we had overcome. More insidiously, it deepens the very polarization it seeks to remedy.

> Since science communication bears the name of communication, it must be an effort towards mutual, deliberative dialogue aimed at realizing a democratic society.—**Mikihito Tanaka**

The harder communicators push facts at skeptical audiences, the more those audiences entrench in opposition. Fundamentally, alarmism breeds distrust among people and sets back our democracy.

The death of the deficit model forces a fundamental reckoning. If the problem is not insufficient information, then what is it? And if the solution is not better explanations, then what should replace them?

Sean Decatur, drawing on research from the Trust Edge Leadership Institute, offers a deceptively simple answer: "Communication is never the core issue when you're trying to deliver a message. Trust is." The insight crystallizes a shift in how we must understand science communication. People struggle to accept ideas from institutions and individuals they do not trust, regardless of how clearly those ideas are conveyed or how robust the evidence behind them. Trust precedes comprehension. Without it, communication efforts cannot gain purchase.

This reframing has profound implications. It means the central challenge is not pedagogical but relational. The question is not "How do we explain this more clearly?" but rather "How do we build relationships that allow explanation to be heard?" Museums like the American Museum of Natural History, which Decatur leads, have long understood this distinction. Their success lies not in the volume of facts they convey but in their capacity to inspire questions.

> If I had to pick the thing that I think is most important for visitors to take away when they come to visit the museum, I'd say they should take away a question. Each diorama, each exhibit, each planetarium show was designed not only to capture attention but to get you to observe and hopefully provoke a question.—**Sean Decatur**

Each exhibit is designed not merely to inform but to provoke curiosity—to draw visitors into the scientific process itself rather than simply presenting its conclusions. This process-oriented approach embodies a crucial insight: trust forms through engagement with *how* science works, not just *what* it has discovered. When visitors see how fossils are found, what methods researchers use, which questions remain open, they develop appreciation for the systematic, self-correcting nature of scientific inquiry. The Titanosaur exhibit at the American Museum draws millions not because it delivers taxonomic facts but because it tells the story of discovery—revealing the messy, iterative, human work behind spectacular findings. The Hall of Human Origins tackles even more sensitive territory, addressing race and human variation precisely where religious and cultural beliefs generate skepticism. Here too, the emphasis rests on evidence: showing genomic data, fossil records, and the process by which researchers construct evolutionary narratives from fragmentary remains.

Yet trust cannot be built through any single institution or communication mode alone. Robbert Dijkgraaf emphasizes that scientists cannot retreat to ivory towers and expect the public to come to them. At the same time, they should not be naive about science communication—effective engagement requires expertise and rigorous application of evidence-based strategies.

Karin Markides extends this argument to structural coordination. When different audiences receive disconnected messages through fragmented channels, society loses its capacity for collective understanding and action. Trust requires not only good relationships between scientists and publics but also coherence across communication efforts. Universities, Markides argues, must serve as "neutral hosts" that convene diverse stakeholders—policymakers, industry leaders, community organizations—to build shared understanding. This coordination function may prove as important as any individual communication act.

Tanaka pushes the argument furthest, demanding that we recognize communication itself as fundamentally dialogic. This is not merely an aesthetic preference or tactical choice—it is a matter of democratic principle. Neither populism (which rejects expertise), nor alarmism (which reverts to deficit thinking), nor accelerationism (which prioritizes speed over deliberation) allows space for genuine dialogue. All three treat publics as passive recipients rather than active participants in shaping how science serves society.

The convergence of these five voices—Nurse, Decatur, Dijkgraaf, Markides, and Tanaka—around trust, process, dialogue, and coordination is striking. Their practical experience across museums, research institutes, science policy organizations, and communication scholarship points toward common conclusions. Yet a crucial question remains: What does the research landscape actually tell us? Where does scholarly attention concentrate? Which domains have matured, and where are the critical gaps between what practitioners identify as essential and what researchers actually study?

To answer these questions, this chapter examines the intellectual landscape of science communication through a bibliometric analysis. This chapter charts that course by integrating the bibliometric landscape with practitioner insights. We examine how the field has evolved in response to crises, where high-impact work concentrates, which promising approaches remain under-researched, and what emerging frontiers signal about future directions.

## 2 Data and Methods

To understand where scholarly attention concentrates within science communication research, we performed a bibliometric analysis using topic modeling. We retrieved 5397 records from Web of Science (December 30, 2025) using the topic search query TS = "science communication". This corpus includes journal articles, conference proceedings, and book chapters spanning multiple decades of research.

We applied transformer-based embeddings to convert titles and abstracts into numeric vector representations, capturing semantic content in high-dimensional space. These embeddings preserve contextual meaning, allowing documents with similar themes to cluster together even when using different terminology. We then employed HDBScan (Hierarchical Density-Based Spatial Clustering of Applications with Noise) to identify cohesive thematic clusters within the embedding space (McInnes et al., 2017). Unlike partition-based methods that force all documents into clusters, HDBScan identifies dense regions of semantically related papers while filtering out peripheral records that do not align with coherent themes.

This process yielded 3453 usable records organized into 48 distinct topical clusters plus one "others" category for documents that did not fit established patterns. Cluster labels were derived from the content and themes of the most highly cited articles within each group, ensuring that names reflect the substantive focus of the research they represent.

For each cluster, we extracted three key characteristics: (1) document count, indicating the volume of research attention; (2) average publication year, revealing temporal positioning and whether a cluster represents established work or emerging scholarship; and (3) average citations per document, serving as a proxy for impact and maturity. Together, these metrics allow us to map the intellectual landscape—identifying well-established domains with substantial evidence bases, emerging frontiers responding to contemporary challenges, and critical gaps where practitioner-identified priorities remain under-researched.

## 3 Results

The bibliometric landscape reveals a field in transformation. Visualizing the 48 clusters by their temporal positioning (publication year), citation impact, and document volume as shown in Fig. 10.1, exposes patterns invisible in traditional literature reviews. Some domains command substantial attention with hundreds of papers and decades of cumulative citations. Others emerge as nascent frontiers, responding to contemporary crises with recent publications still accruing influence. Still others—critically—represent gaps where practitioner priorities far exceed scholarly investigation.

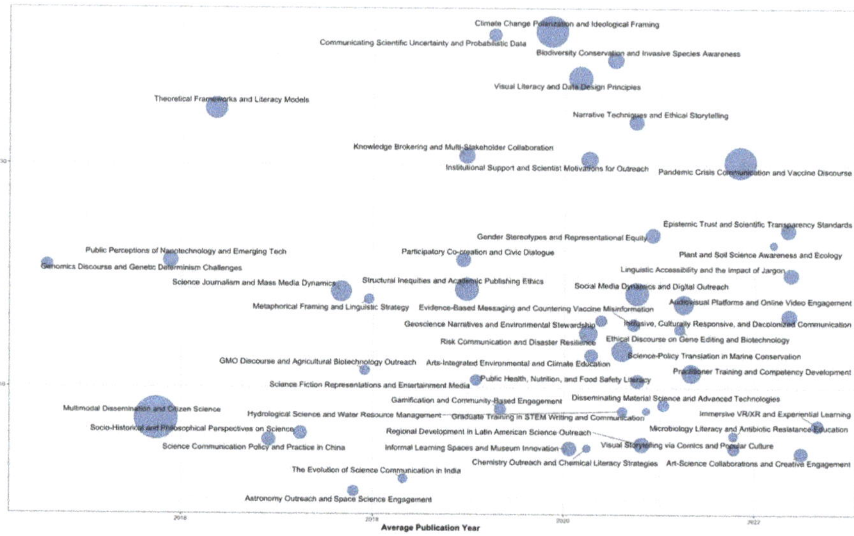


**Fig. 10.1** The science communication landscape mapped across time (x-axis: average publication year) and impact (y-axis: average citations per document). Bubble size represents document count

## *3.1 Temporal Evolution*

The science communication research landscape shows temporal stratification, revealing how the field has responded to successive crises while leaving critical priorities under-examined.

The pre-2020 foundation rests on established theoretical work. Cluster 7 (Theoretical Frameworks and Literacy Models: 94 documents, average publication year 2016.4, 35.0 citations per document) represents the field's conceptual backbone—science literacy frameworks, deficit model critiques, and foundational communication theories. These papers provide essential vocabulary and analytical tools but remain relatively abstract, offering limited practical guidance for practitioners navigating polarized environments.

The largest and highest-impact cluster predates the pandemic but addresses the polarization that would later explode across multiple domains. Cluster 3 (Climate Change Polarization and Ideological Framing) comprises 233 documents averaging 42.0 citations each, with mean publication year 2019.9. This massive, highly cited body of work examines how motivated reasoning, cultural identities, and ideological biases shape climate science reception. The cluster documents why facts alone fail to persuade: identity-protective cognition operates as a filter, causing individuals to process climate information in ways that reinforce rather than challenge their group membership. This work validates what practitioners report—that the deficit model cannot overcome identity-based resistance.

Older clusters reveal the field's foundations. The earliest thematic concentration is Cluster 36 on genomics discourse and genetic determinism challenges (26 documents, 2014.6 average year, 21.0 citations), reflecting early concerns about how media hype and oversimplification distort public understanding of genetic research. Cluster 1 on multimodal dissemination and citizen science is by far the largest cluster in the entire corpus (505 documents, 2015.7 average year), encompassing work on social media strategies, blended learning, and citizen science as a participation mechanism—though its low citation impact (7.2 per document) suggests a broad, practice-oriented base rather than high-impact theoretical contributions. Cluster 16 on public perceptions of nanotechnology and emerging technologies (43 documents, 2015.9 average year, 21.4 citations) represents earlier risk communication scholarship, while Cluster 7 on theoretical frameworks and science literacy models (94 documents, 2016.4 average year, 35.0 citations) provides the conceptual scaffolding. Together, these pre-2020 clusters established both the field's theoretical vocabulary and its practical dissemination infrastructure, but remained focused on models, measurement, and channel-level outreach rather than intervention.

The post-2020 acceleration reveals the field's crisis-driven character. Cluster 2, focusing on pandemic crisis communication and vaccine discourse, exploded into existence with 247 documents averaging 2021.8 publication year—making it the second-largest cluster despite its recency. With 30.2 citations per document, it demonstrates high impact for such recent work. The pandemic forced rapid engagement with misinformation, social media dynamics, preprint culture, and behavioral science insights (Van Bavel et al., 2020; Fraser et al., 2021). Research in this cluster examines the psychological roots of vaccine hesitancy across 24 nations, revealing that conspiratorial thinking and reactance (resistance to perceived threats to freedom) predict anti-vaccination attitudes more strongly than knowledge deficits (Hornsey et al., 2018). This cluster's size and velocity reveal a field scrambling to respond to urgent public health communication needs.

Equally revealing are two clusters that emerged post-2020 and signal profound shifts in how the field conceptualizes its mission. Cluster 21 (Epistemic Trust and Scientific Transparency Standards) appeared in 2022.3 on average, with 38 documents averaging 24.1 citations—evidence of rapid uptake. This cluster treats trust not as a simple attitude but as a distinct theoretical construct, focusing on process transparency, replication standards, and ethical communication practices. Trust, in this emerging view, is separate from agreement or understanding. You can trust scientists' integrity without agreeing with their conclusions. This distinction echoes Decatur's emphasis on process over product: people need to see *how* science works, not just accept *what* it claims.

Similarly, Cluster 18—Inclusive, Culturally Responsive, and Decolonized Communication—represents one of the newest thematic developments (average publication year 2022.4, 42 documents, 16.4 citations). This cluster addresses historical exclusions and advocates for valuing non-Western and indigenous knowledge systems. Nurse's emphasis on cultural sensitivity finds empirical grounding here: the field is beginning to recognize that Western-dominated communication models cannot simply be exported wholesale. Trust-building looks fundamentally different across cultural contexts, and effective communication must honor those differences rather than impose uniform approaches.

The temporal pattern is clear: the field responds to crises rather than anticipating them. Both the pandemic and growing recognition of Western-centric limitations emerged from acute failures—COVID-19 misinformation and the inadequacy of Western frameworks in diverse contexts. This reactive posture leaves communicators perpetually behind the curve, scrambling to address emergencies rather than building resilient trust infrastructure proactively.

## 3.2 High-Impact Domains

The bibliometric landscape reveals four domains where substantial research attention has produced robust evidence bases and high-impact findings. These clusters represent the field's most mature theoretical and empirical work.

Climate Change Polarization and Ideological Framing (Cluster 3) stands as the field's largest of the highest-impact domains, with 233 documents averaging 42.0 citations per paper and a mean publication year of 2019.9. This cluster examines why climate science encounters such fierce resistance despite overwhelming scientific consensus. The research reveals how motivated reasoning, cultural worldviews, and identity-protective cognition create barriers that mere information cannot breach (Kahan et al., 2012; Kahan, 2017, 2013). Work in this domain documents how political identity transforms climate change from a scientific question into a social identity marker, where accepting the consensus risks communal ties (Kahan, 2015). Critically, this research also reveals boomerang effects: Hart and Nisbet's influential study demonstrated how identity cues in science messages can amplify rather than reduce polarization, with Republican participants showing increased opposition after exposure to certain climate communication frames (Hart & Nisbet, 2012). The cluster's size and impact validate what the panelists observed: polarization around science is not a communication failure but a social identity phenomenon requiring different strategies than information provision.

Visual and narrative techniques emerge as a second high-impact domain. Cluster 4 on Visual Literacy and Data Design Principles contains 114 documents (average publication year 2020.2, 37.8 citations per document), demonstrating how strategic use of images, color schemes, and interactive tools can enhance comprehension and engagement (Crameri et al., 2020; Trumbo, 1999). Research in this cluster reveals that scientifically derived color maps reduce complexity while remaining accessible to people with color-vision deficiencies—addressing both accuracy and inclusivity simultaneously. Cluster 20 on Narrative Techniques and Ethical Storytelling (39 documents, 33.9 average citations) complements this, showing how storytelling renders scientific concepts engaging and memorable while promoting behavioral change (Dahlstrom, 2014; ElShafie, 2018). Narratives offer increased comprehension, interest, and engagement precisely because they become self-motivating vehicles for information delivery (Downs, 2014). This dual emphasis on visual and narrative approaches aligns directly with Decatur's museum practice—his exhibits emphasize visual discovery processes and narrative arcs precisely because these modes bypass the cognitive defenses that textual argument alone cannot penetrate.

A third high-impact domain addresses uncertainty communication. Cluster 30 (Communicating Scientific Uncertainty and Probabilistic Data) contains only 30 documents but commands 41.7 citations per paper—the second-highest impact in the entire landscape. This small but influential body of work examines how to present uncertainty transparently without eroding public confidence. The cluster's high impact reflects how challenging and consequential this problem is: audiences need honest assessments of scientific confidence, yet emphasizing uncertainty can be weaponized by those seeking to delay action or sow doubt. The field recognizes this tension but has not yet resolved it satisfactorily.

Biodiversity conservation and invasive species awareness constitutes a fourth high-impact domain. Cluster 14 contains 46 documents averaging 39.4 citations per paper (average publication year 2020.5), making it the third-highest-impact cluster in the landscape. This body of work examines how effective communication about invasive species, pollinator decline, and biodiversity loss can inform policy and mobilize public support for conservation action. The cluster highlights challenges familiar across science communication—science denialism, media biases, and misinformation—but applied to an urgent domain where communication failures translate directly into ecological consequences. Its high citation impact suggests that the intersection of science communication with conservation practice resonates strongly across research communities.

Finally, institutional infrastructure receives sustained attention. Cluster 13 (Institutional Support and Scientist Motivations for Outreach) contains 51 documents averaging 30.5 citations, examining why scientists engage in public communication, what barriers they face, and which policy frameworks facilitate or hinder outreach (Besley et al., 2018; Varner, 2014). Research reveals that the most consistent predictors of scientists' willingness to engage are beliefs about enjoying the experience, making a difference through engagement, and having sufficient time—yet institutional structures rarely support these conditions (Weingart et al., 2021). This work documents a persistent problem: scientists who engage in communication often receive little institutional reward, and in some cases face professional penalties. During the panel discussion, a young scientist raised precisely this concern, asking how to balance research demands with communication expectations. Markides responded that professional science communicators should be "part of the university infrastructure"—a call that the bibliometric evidence shows remains largely unheeded in practice.

## 3.3 *Critical Gaps*

An interesting revelation from the bibliometric landscape concerns not what is present but what is conspicuously absent or dramatically under-represented. The communication strategies that practitioners identify as most promising occupy surprisingly small corners of the research ecosystem, revealing a profound disconnect between evidence needs and scholarly attention.

### 3.3.1 Trusted Messengers

Perhaps no gap is more consequential than the paucity of research on trusted messengers and knowledge brokering. Cluster 15 (Knowledge Brokering and Multi-Stakeholder Collaboration) contains merely 45 documents—representing just 1.3% of the entire corpus. With an average publication year of 2019.0 and 30.8 citations per document, this work is neither cutting-edge nor particularly high-impact despite addressing what may be the most effective communication mechanism available.

The scarcity of research stands in stark contrast to practitioner emphasis. During the STS *forum* panel, Nurse highlighted the effectiveness of local weathercasters in communicating climate science precisely because they "live in communities" and have "skin in the game." Markides called for universities to serve as "neutral hosts" convening trusted local actors across sectors. The logic is compelling: when audiences perceive messengers as sharing their identity or values, skepticism filters relax, allowing information to be heard rather than reflexively rejected. Research within this small cluster recognizes multiple perspectives on knowledge flow: from one-way transmission to two-way collaborative knowledge production, with joint production processes potentially allowing for democratic decision-making (Bertuol-Garcia et al., 2018).

Why does this promising mechanism remain under-studied? Trusted messenger work is inherently local and context-specific—local communicators in one region face different challenges than those in another, religious leaders in evangelical communities operate differently than those in mainline denominations (Wilson et al., 2016). Such context-dependence resists the generalization that academic publishing typically demands. Yet this very specificity may be what makes the approach effective, suggesting that the field's publication norms inadvertently discourage re-search on the most promising practical interventions.

### 3.3.2 Participatory Co-Creation

Cluster 24 (Participatory Co-creation and Civic Dialogue) fares little better, comprising just 37 documents with an average publication year of 2018.9 and 21.5 citations per document. This below-average impact occurs despite mounting calls for moving from broadcast to dialogue, from top-down expert communication to genuine two-way engagement with communities.

This gap may seem surprising given the large body of work on multimodal dissemination and citizen science (Cluster 1: 505 documents), which constitutes the largest cluster in the corpus. However, that cluster focuses primarily on dissemination channels, outreach platforms, and citizen participation in data collection within scientist-defined research frameworks. The deliberative, power-sharing dialogue that characterizes participatory co-creation is qualitatively different: citizen science typically invites publics to contribute within structures that researchers have already established, whereas co-creation asks publics to help define the research questions, priorities, and frameworks themselves. It is this latter form of engagement—genuine two-way negotiation over what science investigates and how it serves society—that remains under-studied.

Tanaka's insistence that science communication "must be an effort towards mutual, deliberative dialogue aimed at realizing a democratic society" captures the normative force behind participatory approaches. Yet if we lack rigorous research on how such dialogue functions, under what conditions it builds trust, which formats prove most effective, and how outcomes can be sustained over time, practitioners must rely on intuition rather than evidence.

The research gap likely reflects the methodological challenges participatory work presents. Such projects are slow, resource-intensive, and produce outcomes that resist easy quantification. They require sustained community partnerships that don't fit grant cycles. The "effects" are diffuse—shifts in relationships, changes in institutional responsiveness, gradual trust accumulation—rather than discrete, measurable endpoints. Academic incentive structures penalize this kind of research, even as practitioners increasingly recognize its importance.

### 3.3.3 Multi-Stakeholder Coordination

No distinct cluster addresses multi-stakeholder coordination as a research focus. While coordination appears in descriptions of other clusters, it hasn't generated sufficient scholarly attention to constitute its own thematic domain. This absence is remarkable given the panel's convergence on coordination as essential infrastructure for effective science communication at scale.

> Information silos really are the enemy of transformative impact. There needs to be society-wide trust in verified science to enable progress.—**Karin Markides**

Markides warned that "information silos really are the enemy of transformative impact," arguing for coordinated efforts across universities, media, policymakers, and community organizations. Dijkgraaf emphasized "transdisciplinary" work requiring partnerships between scientists and communication professionals. Yet the bibliometric landscape reveals almost no research examining how such coordination occurs, what institutional structures enable it, which governance mechanisms sustain it, or how effectiveness can be evaluated.

The gap exists because coordination research demands partnering across institutional sectors that rarely collaborate—universities, media organizations, government agencies, community groups. Such work falls between disciplinary boundaries, fitting neatly into no single field's jurisdiction. Funding mechanisms typically support single-sector research, not multi-stakeholder infrastructure studies. As a result everyone agrees coordination matters, but no one studies it systematically.

## 3.4 *Emerging Frontiers*

Despite these critical gaps, the bibliometric landscape reveals promising developments in four frontier areas that may reshape how science communication is understood and practiced.

### 3.4.1 Epistemic Trust

Cluster 21 (Epistemic Trust and Scientific Transparency Standards) represents one of the most significant recent conceptual developments. With 38 documents, an average publication year of 2022.3, and 24.1 citations per document, this emerging cluster demonstrates both recency and impact. The work marks a maturation beyond simplistic notions that trust equals agreement or understanding.

This frontier explores trust as a distinct theoretical construct—one that can exist even in the absence of comprehension or consensus (Hendriks et al., 2016). The research examines how process transparency builds epistemic trust: open data practices, pre-registration of studies, disclosure of funding sources, public acknowledgment of uncertainty and limitations. Research demonstrates that critical characteristics determining epistemic trustworthiness include not only expertise but also integrity and benevolence—dimensions that must be communicated through visible process (Besley et al., 2021). Crucially, these practices increase trust even among audiences who lack technical capacity to audit the work themselves. The visibility of rigorous process serves as a trust signal (John, 2018).

Dijkgraaf's notion of "truth in advertising"—delivering real solutions while acknowledging limitations—captures this frontier's practical implications. So too does the finding that 92% of the public values scientists "being open to changing their minds" in light of new evidence. This isn't about getting the answer right the first time; it's about demonstrating a reliable process for self-correction.

The frontier suggests that science communication might focus less on explaining findings and more on revealing process—showing how science works, what counts as evidence, why conclusions change, where uncertainty remains. This process-oriented approach aligns with Decatur's museum philosophy but extends it through systematic research on mechanisms and effects.

### 3.4.2 Inclusive and Decolonized Communication

Cluster 18 (Inclusive, Culturally Responsive, and Decolonized Communication) is among the newest thematic developments in the landscape, with an average publication year of 2022.4 and 42 documents generating 16.4 citations each. Though impact remains modest as the frontier establishes itself, the cluster's emergence signals growing recognition that Western-dominated communication models are inadequate for global science.

This work addresses historical exclusions—how science communication has privileged certain languages, cultural frameworks, and ways of knowing while marginalizing others (Dawson, 2018; Canfield et al., 2020). It examines power dynamics: who gets to define what counts as "science," which publics are deemed worthy of engagement, whose concerns receive serious attention, and how colonial legacies continue shaping communication practices (Finlay et al., 2021). Research reveals that participation in science communication is marked by structural inequalities, particularly ethnicity and class, with marginalized groups experiencing cultural imperialism and powerlessness in their encounters with science communication.

Nurse's observation that dominant science nations often "talk to other cultures using language that is appropriate to their own country and not the country they are speaking to" identifies the problem this frontier tackles. The research explores culturally responsive approaches that honor local knowledge systems, engage indigenous ways of knowing, and recognize that trust-building differs profoundly across cultural contexts.

This frontier remains under-developed—42 documents is a small base for a challenge this significant. Yet its emergence as a distinct cluster suggests the field is beginning to confront uncomfortable questions about whose science communication matters and whose publics count.

### 3.4.3 Immersive Technologies

Cluster 32 (Immersive VR/XR and Experiential Learning) occupies the cutting edge temporally, with an average publication year of 2022.7, but remains nascent in both size (29 documents) and impact (6.6 citations per document). This frontier explores virtual and extended reality technologies for science engagement—immersive experiences that could transform how publics encounter scientific concepts.

The research examines VR applications for environmental communication, allowing users to experience climate change impacts viscerally rather than abstractly. It tests immersive learning for complex scientific processes, evaluating whether experiential engagement increases understanding beyond traditional media. Early findings suggest potential for affective engagement that textual or even visual communication cannot match.

Yet this frontier embodies precisely the tension Tanaka identifies in his "accelerationism" critique: technology advancing faster than trust-building can occur. The tools may be powerful, but if deployed without addressing underlying distrust, they risk amplifying rather than resolving communication failures. The frontier's low impact suggests the field is still determining whether immersive technologies offer genuine solutions or merely novel gimmicks.

### 3.5 *Non-Western Contexts*

Three small clusters reveal how Western-dominated the science communication landscape remains. Cluster 25 (Science Communication Policy and Practice in China) contains 34 documents from 2016.9 averaging just 5.3 citations. Cluster 45 (The Evolution of Science Communication in India) includes 22 documents from 2018.3 with a mere 1.8 citations per document. Cluster 17 (Regional Development in Latin American Science Outreach) encompasses 42 documents from 2020.8 averaging 4.9 citations.

Together, these 98 documents—representing China, India, and Latin America—constitute less than 3% of the corpus despite these regions comprising the majority of the world's population. The low citation counts suggest this work remains marginalized within the field's attention economy, rarely cited by researchers working in Western contexts.

Nurse's warning finds empirical validation: the science communication literature reflects and reproduces the global North's dominance. If trust-building requires cultural responsiveness, as the inclusive communication frontier argues, then this severe underrepresentation of non-Western contexts represents not merely an equity issue but a pragmatic failure. The field cannot develop globally applicable approaches while studying almost exclusively Western contexts.

## 4 Discussion

The bibliometric landscape offers both validation and indictment. On one hand, it confirms what the panelists observed: polarization mechanisms are well-understood, visual and narrative approaches rest on solid evidence bases, and the field has begun grappling with epistemic trust as a distinct construct. Research reveals why people resist scientific consensus—identity-protective cognition, motivated reasoning, and cultural worldview defense operate as powerful filters that information alone cannot breach (Kahan et al., 2012; Kahan, 2017). The mechanisms that produce polarization have been exhaustively mapped.

On the other hand, the landscape exposes a troubling mismatch between what practitioners identify as most effective and where scholarly attention concentrates. The trust-building strategies that Nurse, Decatur, Dijkgraaf, Markides, and Tanaka converge upon—trusted messengers, participatory co-creation, multi-stakeholder coordination—occupy the smallest corners of the research ecosystem. The small knowledge brokering cluster stands out as one of the most promising mechanisms for bridging identity divides yet remains dramatically under-studied. Participatory approaches receive minimal scholarly attention despite mounting calls for moving from broadcast to dialogue. Multi-stakeholder coordination—which all five panelists emphasized—has generated insufficient research to constitute its own cluster.

This gap explains why well-intentioned interventions keep failing. When researchers design communication strategies based on what the literature emphasizes rather than what practice reveals to be effective, they produce interventions optimized for the wrong outcomes. Brief messaging experiments proliferate because they are methodologically tractable and fit publication norms, even as evidence accumulates that they fail to build durable trust. As discussed in the Introduction, the most rigorous test to date—Gligorić and colleagues' large-scale study—found zero effect across all interventions (Gligorić et al., 2025).

Recent work elaborates why: trust reflects stable attitudes resistant to simple messaging fixes (Wintterlin, 2025). Trust formation requires extended engagement, repeated positive interactions, and visible alignment between messenger values and audience values. Yet the research infrastructure rewards quick experiments over sustained relationship-building, measurable outcomes over gradual trust accumulation, and generalizable findings over context-specific adaptation.

Interventions succeeding in building trust operated through mechanisms the bibliometric landscape reveals to be under-studied (Nisbet et al., 2015; Post & Bienzeisler, 2024; Fiske & Dupree, 2014). The honest broker approach—where scientists separate empirical findings from policy recommendations—attenuates polarization precisely because it reduces identity threat (Post & Bienzeisler, 2024). By not forcing audiences to choose between scientific consensus and their political values, the approach provides an off-ramp from defensive rejection. Yet this strategy requires restraint that runs counter to how scientists often engage: presenting findings while explicitly advocating for specific policies.

Consensus messaging demonstrates the importance of context-dependence that the literature underemphasizes. Dunwoody and colleagues found that weight-of-experts statements increased certainty and acceptance of scientific judgments (Dunwoody & Kohl, 2017). However, Chinn and colleagues revealed this effect is moderated by baseline trust: consensus messaging only works when audiences already perceive the scientific community as credible (Chinn et al., 2018). For audiences with low trust—precisely those most in need of persuasion—highlighting consensus can backfire, triggering reactance rather than acceptance. This conditional effectiveness demands the audience segmentation that bibliometric analysis shows receives insufficient attention.

The integration becomes clear: research landscape confirms the trust crisis and polarization mechanisms are well-understood. Visual and narrative techniques have robust evidence bases, uncertainty communication has generated influential work, and epistemic trust is emerging as a promising frontier. But the solutions practitioners identify—trusted messengers, long-term engagement, multi-stakeholder coordination—are precisely where research is thinnest. This mismatch explains why well-intentioned interventions keep failing and points toward what a more effective research agenda would prioritize.

Moving from diagnosis to prescription requires a framework that honors both the evidence base and the gaps it reveals. The panelists converge on several principles: match strategies to specific audiences (Nurse's cultural sensitivity), operate through validated mechanisms (Dijkgraaf's "science of science communication"), build relational trust over time (Decatur's primacy of trust), coordinate across stakeholders (Markides' anti-silo vision), and enable dialogue not broadcast (Tanaka's democratic imperative). Integrating these with the academic landscape of the field suggests the presence of a two-dimensional framework organized around audience segmentation, and mechanism pathways.

## *4.1 Dimension 1: Audience Segmentation*

Effective science communication must differentiate strategies based on audience characteristics, particularly their baseline trust levels and the nature of their skepticism. The literature and practice converge on three broad segments requiring distinct approaches.

**Segment A: The Persuadable Middle** Characterized by moderate trust, uncertainty, and openness to clear signals—responds well to consensus messaging. Chinn and colleagues demonstrated that communicating higher levels of scientific agreement increases perceived certainty, which in turn predicts greater personal agreement and policy support for non-politicized issues (Chinn et al., 2018). Critically, this effect operates through trust as a moderator: those with moderate-to-high baseline trust use consensus information as a reliable heuristic for truth. Dunwoody's findings that weight-of-experts statements increase confidence in scientific judgments operate through the same mechanism (Dunwoody & Kohl, 2017).

Decatur's museum visitors largely occupy this segment—they arrive with sufficient curiosity and trust to engage, making them receptive to well-crafted exhibits that provoke questions rather than demand agreement. The bibliometric landscape confirms this is well-studied territory: the climate polarization cluster (233 documents) and theoretical frameworks cluster (94 documents) provide robust foundations for understanding how this audience processes information.

However, there are limitations. Consensus messaging fails completely for audiences with low baseline trust, and can even increase polarization when perceived as coercive (Gligorić et al., 2025). The segment's boundaries matter: move too far toward skepticism, and entirely different strategies become necessary.

**Segment B: The Identity-Conflicted** Individuals whose political or cultural values clash with scientific consensus—require approaches that reduce identity threat rather than intensify it. Post and colleagues demonstrated this through a crucial experimental design: presenting scientists as "honest brokers" who acknowledge the limitations of their perspective and present policy options, rather than as "epistocrats" who blur scientific and political claims (Post & Bienzeisler, 2024). The honest broker framing attenuated political polarization of trust in scientists, with effects particularly strong among "the most politically challenged"—those whose policy preferences conflicted most with implied recommendations.

The mechanism is clear: when audiences perceive that accepting scientific findings will force them to adopt specific policies that violate their values, they defensively reject the science itself to protect their identity (Kahan et al., 2012). The honest broker approach prevents this contamination by explicitly separating descriptive claims from normative ones, allowing audiences to accept empirical findings while maintaining their own value-based policy preferences.

Visual strategies offer another pathway for this segment. Bolsen and colleagues found that visual imagery could counteract science politicization effects where text-only approaches failed (Bolsen et al., 2019). The combination of textual frames with visual evidence—animated maps showing environmental changes over time—significantly increased belief that climate change is happening and will have negative impacts. Visual information engages dual-coding pathways and appears less susceptible to motivated reasoning than purely verbal arguments, providing concrete anchors that are harder to dismiss through abstract counterarguments.

Moral reframing represents a third evidence-based approach for the identity-conflicted. Research from multiple studies demonstrates that aligning scientific messages with conservative moral foundations—purity, authority, loyalty—increases acceptance among audiences who would reject the same message framed through liberal foundations like care and fairness. When climate action is framed as "patriotic energy independence" or "stewardship of God's creation" rather than "protecting vulnerable populations" or "ensuring intergenerational justice," conservative audiences show markedly higher support.

> Scientists are part of society. They are normal people—well, most of the time, normal sorts of people… sometimes the wearing of white coats on television just puts a barrier between the scientists and the individuals they're trying to speak to.—**Paul Nurse**

Nurse's insight about removing these "white coat" barriers operationalizes this principle: presenting science without the cultural baggage that triggers identity defense. The bibliometric landscape confirms this segment receives substantial attention—the climate polarization cluster is the largest and highest-impact domain—but solutions remain works in progress, with visual and narrative clusters offering promising approaches.

**Segment C: The Structurally Distrustful** Those with fundamental institutional skepticism—present the greatest challenge and receive the least adequate research attention. Gligorić's null findings are definitive: brief interventions targeting this segment simply do not work (Gligorić et al., 2025). Five theoretically grounded messaging strategies, carefully designed to address known mechanisms of distrust, produced zero measurable effect among conservatives with low baseline trust in scientists.

What does work, according to limited evidence, is extended epistemological literacy. Viciana and colleagues found that teaching individuals *how to evaluate knowledge claims* rather than *what to believe* produced a 14% increase in belief updating toward scientific consensus after 2 months (Viciana et al., 2024). Crucially, this effect held across both ideologically congruent and incongruent scientific results, suggesting the mechanism operates at a meta-cognitive level rather than through persuasion about specific issues.

The intervention provides tools for adjudicating between competing claims—understanding peer review, recognizing replication as a validity signal, distinguishing correlation from causation—creating a generalizable framework that audiences can apply independently. This depth requires time that brief interventions cannot provide, explaining why surface-level messaging fails while sustained engagement succeeds.

Tanaka's distinction becomes relevant here: he differentiates moderate conspiracy believers—who can be reached through appropriate strategies—from extreme conspiracy theorists whose rage is directed toward society rather than science per se, requiring social support rather than science communication. The structurally distrustful segment occupies this middle ground: deep skepticism that brief messaging cannot overcome but that sustained epistemological work can address.

## 4.2 Dimension 2: Mechanism Pathways

Understanding audience segments clarifies *whom* to reach; mechanism pathways address *how* trust is built and maintained. Four pathways emerge from converging evidence across bibliometric analysis and practitioner experience.

**Pathway 1: Relational Trust Transfer (Trusted Messengers)** Operates through in-group identity to bypass skepticism filters. Altenmüller and colleagues demonstrated the power of perceived shared identity: when scientists were perceived as conservative, conservative audiences' trust increased significantly (Altenmüller et al., 2024). The mechanism is straightforward: in-group messengers receive a presumption of shared values and interests, reducing the identity threat that out-group messengers automatically trigger.

Prior research evidence validates the approach across domains. The Climate Matters program trained 1100 local weathercasters across 94% of U.S. media markets, resulting in more than 100-fold increase in climate reporting. Randomized experiments confirm that viewers of these localized reports show significantly higher understanding that climate change is happening, is human-caused, and is locally relevant—effects that hold even when controlling for political ideology (Maibach et al., 2020). A conservative viewer distrusts the IPCC but trusts their local weathercaster, who lives in their community and has "skin in the game."

Similarly, the Science for Seminaries program equips clergy to engage scientific topics within theological contexts, creating permission structures for congregants to accept science without compromising faith. When a pastor explains why environmental stewardship is a biblical mandate supported by climate science, parishioners can integrate both value systems rather than choosing between them.

Markides' framing of universities as "neutral hosts" convening trusted local actors operationalizes this pathway at institutional scale. The mechanism succeeds because trust transfers through relational bonds that precede and outlast any specific message.

Research priorities should aim for systematic study of messenger characteristics that maximize trust transfer, scaling strategies for messenger networks, mechanisms for maintaining message fidelity while allowing local adaptation, and cross-domain testing to identify which messenger types work for which audiences and issues.

**Pathway 2: Cognitive Frameworks (Epistemological Literacy)** Operates by teaching meta-knowledge about evaluating claims rather than asserting specific conclusions. Viciana's findings demonstrate the power of this approach: individuals trained in scientific epistemology showed 14% higher belief updating toward consensus after 2 months, with effects robust across ideologically congruent and incongruent issues (Viciana et al., 2024). Higher epistemological literacy predicted refusal of pseudoscientific beliefs, higher scores on methodology tests, and reduced epistemic mistrust.

The mechanism operates at a different level than persuasion. Rather than arguing "climate change is real," epistemological literacy teaches: "Here's how to recognize a well-designed study. Here's why replication matters. Here's what peer review checks for and what it misses. Here's how to evaluate whether a source has relevant expertise." Audiences equipped with these tools can apply them independently across domains, building durable resistance to misinformation.

> Scientists have to become comfortable being uncomfortable and make sure that the base in society gets larger. That means we should definitely not retreat to our ivory towers, but we should also not be naive about science communication… There is a field called the science of science communication.—**Robbert Dijkgraaf**

Dijkgraaf's insistence that "there is a field called the science of science communication" captures this pathway's foundation: scientific study has identified what works, and communicators must apply those findings rather than relying on intuition. The bibliometric landscape reveals strong theoretical foundations (cluster 7 on frameworks and literacy models) but limited application research. translating theoretical understanding into practical programs that can be delivered at scale, with careful evaluation of which pedagogical approaches work for which populations remains as a gap.

**Pathway 3: Affective Engagement (Visual Narrative)** Operates through storytelling and imagery to engage processing pathways that bypass motivated reasoning. Bolsen's work on visual imagery demonstrates how data visualizations—animated maps showing temperature changes, satellite images revealing ice loss—make abstract claims concrete and harder to dismiss (Bolsen et al., 2019). The mechanism likely involves grounding: visual evidence provides perceptual anchors that reduce the cognitive space available for counter-arguing.

Narrative approaches complement this. Dahlstrom and colleagues found that narrative formats increase comprehension, interest, and engagement compared to expository presentations (Dahlstrom, 2014). Stories become self-motivating vehicles for information delivery, drawing audiences into content they might otherwise resist. The key is presenting the *process* of discovery—how scientists found evidence, what questions arose, which hypotheses failed—rather than merely stating conclusions.

Decatur's museum practice, described in the Introduction, exemplifies this pathway—exhibits succeed not by asserting conclusions but by immersing visitors in the process of discovery, engaging precisely the affective and narrative pathways that research identifies as effective.

The landscape confirms robust foundations: Cluster 4 on visual literacy and Cluster 20 on narrative techniques demonstrate this pathway's maturity. However, visual and narrative approaches work best for moderately skeptical audiences, proving less effective for the deeply distrustful who resist any scientific message regardless of format.

**Pathway 4: Structural Transparency (Open Science)** Operates by making the scientific process visible, reducing the space where conspiracy theories can flourish. Lupia and colleagues found that 84% of Americans consider it important for scientists to disclose their funders, while 92% value scientists "being open to changing their minds based on new evidence" (Lupia et al., 2024). The public doesn't expect infallibility; they expect honesty about uncertainty, acknowledgment of limitations, and transparent correction when errors occur.

Open science practices—preregistration of studies, open data, disclosed funding sources—serve as trust signals even for audiences lacking technical capacity to audit the work themselves (Nosek et al., 2015). The visibility of rigorous process communicates integrity and benevolence, two critical dimensions of trustworthiness beyond mere expertise. When scientists use open practices and publicly disclose them, audiences perceive reduced bias and increased reliability.

The emerging epistemic trust cluster (38 documents, 2022.3 average year, 24.1 citations) represents the cutting edge of this research frontier. Work in this domain treats trust as distinct from agreement or understanding—you can trust scientists' integrity without accepting their conclusions. This distinction echoes Decatur's emphasis: the museum succeeds not by demanding visitors accept specific claims but by revealing *how scientists know what they claim to know*.

Dijkgraaf's "truth in advertising" principle operationalizes this: delivering real solutions while acknowledging what remains uncertain, what we got wrong before, what trade-offs different courses of action involve. Tanaka's emphasis on transparency as essential for "mutual, deliberative dialogue" extends the argument: democracy requires that citizens can see and evaluate the reasoning behind claims that affect their lives.

The pathway shows promise but requires cultural change within science. Many researchers resist open practices, fearing they invite bad-faith criticism or create competitive disadvantages. Institutions must incentivize transparency through hiring, promotion, and funding decisions—treating openness as a professional responsibility rather than optional virtue.

## 5 Conclusion

The science communication landscape of 3453 papers reveals a field that understands polarization deeply but under-invests in the mechanisms practitioners identify as most effective. We have 233 documents explaining why people resist science, but only 45 examining how trusted messengers build bridges across divides. We study brief interventions that consistently fail while neglecting the sustained relationship-building that succeeds. The mismatch between research attention and practical effectiveness explains why well-intentioned communication efforts continue to disappoint.

Yet the path forward is clear. Five voices from the STS *forum* panel—representing museums, research institutes, universities, and communication scholarship—converge on a vision: science communication must transform from broadcast to dialogue, from deficit thinking to trust-building, from isolated interventions to coordinated engagement. Nurse reminds us that science and democracy share foundations in evidence, argument, and truth-seeking. Decatur demonstrates that trust precedes comprehension. Dijkgraaf insists we must become comfortable being uncomfortable while applying the science of science communication. Markides calls for breaking down information silos through coordination. Tanaka demands mutual deliberative dialogue over enlightenment campaigns.

This vision aligns with where evidence is strongest—relational trust through local messengers, extended epistemological literacy, visual and narrative engagement, structural transparency—and where research investment is weakest. The framework presented here offers a diagnostic approach: segment audiences by trust levels, match mechanisms to contexts, align timescales with depth of distrust.

The challenge is formidable but not insurmountable. For researchers, it means studying messy, context-dependent work that doesn't fit neat experimental designs—trusted messenger networks, participatory co-creation, multi-stakeholder coordination. For funders, it requires reallocating resources from failed brief interventions toward sustained engagement programs. For practitioners, it demands applying existing evidence while documenting local successes to build cumulative knowledge.

Tanaka observed that "true, valuable science communication is being tested now more than ever." The test is whether we can move beyond the deficit model's comfortable assumptions toward the harder work of relationship-building across difference. The bibliometric evidence shows we have the knowledge. The panel shows we have the wisdom. What remains is institutional will—to realign incentives, coordinate stakeholders, and invest in trust as the democratic infrastructure it truly is. When science and democracy both honor truth, evidence, and reasoned debate, both flourish. That mutual flourishing is not only possible but essential, and the tools to achieve it are finally coming into focus.

**Competing Interests** No competing interests.

## Session Panelists

**Nurse, Paul**
Honorary Life President, The Francis Crick Institute, U.K.; Secretary General, EMBO [Nobel Laureate 2001 (Physiology or Medicine)]
Sir Paul Maxime Nurse is a geneticist, President of the Royal Society and former Chief Executive and Director of the Francis Crick Institute. He was awarded the 2001 Nobel Prize in Physiology or Medicine, along with Leland Hartwell and Tim Hunt, for their discoveries of protein molecules that control the division of cells in the cell cycle.

**Decatur, Sean**
President, Office of the President, American Museum of Natural History, U.S.A.
Sean M. Decatur became President of the American Museum of Natural History in 2023 after nearly a decade as president of Kenyon College. He previously served as dean of the College of Arts and Sciences and professor of chemistry and biochemistry at Oberlin College, professor at Mount Holyoke College, and visiting scientist at MIT. He is a Fellow of the American Association for the Advancement of Science, a member of the American Academy of Arts and Sciences, and serves on the boards of Swarthmore College, the Joyce Foundation, the Research Corporation for Science Advancement, New York City Tourism +Conventions, and the Bank Street College of Education. He earned his bachelor's degree from Swarthmore College and his Ph.D. degree in biophysical chemistry from Stanford University.

**Dijkgraaf, Robbert**
President-Elect, International Science Council (ISC), France; Distinguished University Professor, University of Amsterdam, Netherlands
Robbert Dijkgraaf is President-Elect of the International Science Council, the most comprehensive organization representing the global science community, and a Distinguished University Professor at the University of Amsterdam. He is a mathematical physicist who recently served as Minister of Education, Culture and Science of the Netherlands. Before that he was for ten years Director and Leon Levy Professor at the Institute for Advanced Study in Princeton. He is an experienced public policy adviser having served as President of the Royal Netherlands Academy of Arts and Sciences and the InterAcademy Partnership, the global alliance of science academies. For his contributions to science, he has received the Spinoza Prize, the highest scientific award in the Netherlands.

**Markides, Karin**
President and CEO, Okinawa Institute of Science and Technology (OIST), Japan
Dr. Karin Markides earned her doctorate degree from University of Stockholm in Sweden and specializes in analytical chemistry. After serving as President and CEO of Chalmers University of Technology in Sweden as well as American University of Armenia, she became President and CEO of Okinawa Institute of Science and Technology on June 1, 2023. Dr. Markides is appointed Professor Emeriti of Analytical Chemistry at Uppsala University, Elected Member of the Royal Swedish Academy of Engineering Sciences (IVA) and the Royal Swedish Academy of Sciences (KVA), where she has been involved in the selection of the Nobel Prize in Chemistry and the promotion of science.

**Tanaka, Mikihito**
Professor, Faculty of Political Science and Economics, Waseda University, Japan
Mikihito Tanaka is a Professor of Science and Media Studies at the Faculty of Political Science and Economics, Waseda University, Japan. He earned his Ph.D. from the University of Tokyo and has experience as a journalist in parallel to his academic career. Currently, he conducts research on issues between science and society, social media, and journalism. He is a founding member and research manager of the Science Media Centre of Japan (SMCJ). During the COVID-19 pandemic, Tanaka was assigned as a member of the COVID-19 Expert Advisory Board in the Ministry of Health, Labor and Welfare and a risk communication advisor for the Tokyo Metropolitan Government.

## References

Altenmüller, M., Tobias Wingen, and Anna Schulte. 2024. "Explaining Polarized Trust in Scientists: A Political Stereotype-Approach." *Science Communication* 46 (1): 3–36. https://doi.org/10.1177/10755470231221770.

Bertuol-Garcia, Daniella, Carla Morsello, Charbel N. El-Hani, and Renata Pardini. 2018. "A Conceptual Framework for Understanding the Perspectives on the Causes of the Science-Practice Gap in Ecology and Conservation." *Biological Reviews* 93 (2): 1032–55. https://doi.org/10.1111/brv.12385.

Besley, John C., Anthony Dudo, Shupei Yuan, and Forrest Lawrence. 2018. "Understanding Scientists' Willingness to Engage." *Science Communication* 40 (5): 559–90. https://doi.org/10.1177/1075547018786561.

Besley, John C., Nicole I. M. Lee, and Geah Pressgrove. 2021. "Reassessing the Variables Used to Measure Public Perceptions of Scientists." *Science Communication* 43 (1): 3–32. https://doi.org/10.1177/1075547020949547.

Bolsen, T., R. Palm, and J. Kingsland. 2019. "Counteracting Climate Science Politicization with Effective Frames and Imagery." *Science Communication* 41 (2): 147–71. https://doi.org/10.1177/1075547019834565.

Canfield, Kathleen N., Sunshine Menezes, Stephanie B. Matsuda, et al. 2020. "Science Communication Demands a Critical Approach That Centers Inclusion, Equity, and Intersectionality." *Frontiers in Communication* 5: 2. https://doi.org/10.3389/fcomm.2020.00002.

Chinn, Sedona, D. Lane, and P. S. Hart. 2018. "In Consensus We Trust? Persuasive Effects of Scientific Consensus Communication." *Public Understanding of Science* 27 (7): 807–23. https://doi.org/10.1177/0963662518791094.

Crameri, Fabio, Grace E. Shephard, and Philip J. Heron. 2020. "The Misuse of Colour in Science Communication." *Nature Communications* 11: 5444. https://doi.org/10.1038/s41467-020-19160-7.

Dahlstrom, Michael F. 2014. "Using Narratives and Storytelling to Communicate Science with Nonexpert Audiences." *Proceedings of the National Academy of Sciences* 111 (Supplement 4): 13614–20. https://doi.org/10.1073/pnas.1320645111.

Dawson, Emily. 2018. "Reimagining Publics and (Non) Participation: Exploring Exclusion from Science Communication Through the Experiences of Low-Income, Minority Ethnic Groups." *Public Understanding of Science* 27 (7): 772–86. https://doi.org/10.1177/0963662517750072.

Downs, Julie S. 2014. "Prescriptive Scientific Narratives for Communicating Usable Science." *Proceedings of the National Academy of Sciences* 111 (Supplement 4): 13627–33. https://doi.org/10.1073/pnas.1317502111.

Dunwoody, S., and P. Kohl. 2017. "Using Weight-of-Experts Messaging to Communicate Accurately about Contested Science." *Science Communication*. https://doi.org/10.1177/1075547017707765.

ElShafie, Sarah J. 2018. "Making Science Meaningful for Broad Audiences Through Stories." *Integrative and Comparative Biology* 58 (6): 1213–23. https://doi.org/10.1093/icb/icy103.

Finlay, Sonia M., Sujatha Raman, Elizabeth Rasekoala, et al. 2021. "From the Margins to the Mainstream: Deconstructing Science Communication as a White, Western Paradigm." *JCOM-Journal of Science Communication* 20 (1): C03. https://doi.org/10.22323/2.20010302.

Fiske, S., and Cydney H. Dupree. 2014. "Gaining Trust as Well as Respect in Communicating to Motivated Audiences about Science Topics." *Proceedings of the National Academy of Sciences* 111: 13593–97. https://doi.org/10.1073/pnas.1317505111.

Fraser, Nicholas, Liam Brierley, Gautam Dey, Jessica K. Polka, Máté Pálfy, Federico Nanni, and Jonathon Alexis Coates. 2021. "The Evolving Role of Preprints in the Dissemination of COVID-19 Research and Their Impact on the Science Communication Landscape." *PLOS Biology* 19 (4): e3000959. https://doi.org/10.1371/journal.pbio.3000959.

Funk, Cary, and Brian Kennedy. 2024. "Partisan Divides in Trust in Scientists Remain Large." Pew Research Center.

Gligorić, Vukašin, Gerben A. van Kleef, and B. Rutjens. 2025. "Political Ideology and Trust in Scientists in the USA." *Nature Human Behaviour*. https://doi.org/10.1038/s41562-025-02147-z.

Hart, P. Sol, and Erik C. Nisbet. 2012. "Boomerang Effects in Science Communication: How Motivated Reasoning and Identity Cues Amplify Opinion Polarization about Climate Mitigation Policies." *Communication Research* 39 (6): 701–23. https://doi.org/10.1177/0093650211416646.

Hendriks, Friederike, Dorothe Kienhues, and Rainer Bromme. 2016. "Trust in Science and the Science of Trust," 143–59. https://doi.org/10.1007/978-3-319-28059-2_8.

Hornsey, Matthew J., Emily A. Harris, and Kelly S. Fielding. 2018. "The Psychological Roots of Anti-Vaccination Attitudes: A 24-Nation Investigation." *Health Psychology* 37 (4): 307–15. https://doi.org/10.1037/hea0000586.

John, Stephen. 2018. "Epistemic Trust and the Ethics of Science Communication: Against Transparency, Openness, Sincerity and Honesty." *Social Epistemology* 32 (2): 75–87. https://doi.org/10.1080/02691728.2017.1410864.

Kahan, Dan M. 2013. "Ideology, Motivated Reasoning, and Cognitive Reflection." *Judgment and Decision Making* 8: 407–24.

Kahan, Dan M. 2015. "Climate-Science Communication and the Measurement Problem." *Political Psychology* 36: 1–43. https://doi.org/10.1111/pops.12244.

Kahan, Dan M. 2017. "The Expressive Rationality of Inaccurate Perceptions." *Behavioral and Brain Sciences* 40: e6. https://doi.org/10.1017/S0140525X15002332.

Kahan, Dan M., Ellen Peters, Maggie Wittlin, Paul Slovic, Lisa Larrimore Ouellette, Donald Braman, and Gregory Mandel. 2012. "The Polarizing Impact of Science Literacy and Numeracy on Perceived Climate Change Risks." *Nature Climate Change* 2: 732–35. https://doi.org/10.1038/nclimate1547.

Lupia, A., David B. Allison, K. Jamieson, Jennifer Heimberg, Magdalena Skipper, and Susan M. Wolf. 2024. "Trends in US Public Confidence in Science and Opportunities for Progress." *Proceedings of the National Academy of Sciences* 121 (6): e2319488121. https://doi.org/10.1073/pnas.2319488121.

Maibach, Edward, Danielle Perkins, Zhanna Francis, Teresa Myers, David Englert, and Xiaoquan Zhao. 2020. "Localized Climate Reporting by TV Weathercasters Enhances Public Understanding of Climate Change as a Local Problem: Evidence from a Randomized Controlled Experiment." *Bulletin of the American Meteorological Society* 101 (7): E1092–1100. https://doi.org/10.1175/BAMS-D-19-0079.1.

McInnes, Leland, John Healy, and Steve Astels. 2017. "Hdbscan: Hierarchical Density Based Clustering." In *Journal of Open Source Software*, 2:205. 11. https://doi.org/10.21105/joss.00205.

National Academy of Sciences. 2024. "Understanding and Addressing Trust in Science." National Academies Press.

Nisbet, E., Kathryn E. Cooper, and R. K. Garrett. 2015. "The Partisan Brain." *The ANNALS of the American Academy of Political and Social Science* 658 (1): 36–66. https://doi.org/10.1177/0002716214555474.

Nosek, Brian A., George Alter, George C. Banks, Denny Borsboom, Sara D. Bowman, Steven J. Breckler, Stuart Buck, et al. 2015. "Promoting an Open Research Culture." *Science* 348 (6242): 1422–25. https://doi.org/10.1126/science.aab2374.

Pew Research Center. 2023. "Americans' Trust in Scientists, Positive Views of Science Continue to Decline." https://www.pewresearch.org/science/.

Post, Senja, and Nils Bienzeisler. 2024. "The Honest Broker Versus the Epistocrat: Attenuating Distrust in Science by Disentangling Science from Politics." *Political Communication*. https://doi.org/10.1080/10584609.2024.2317274.

Trumbo, Jean. 1999. "Visual Literacy and Science Communication." *Science Communication* 20 (4): 409–25. https://doi.org/10.1177/1075547099020004004.

Van Bavel, Jay J., Katherine Baicker, Paulo S. Boggio, et al. 2020. "Using Social and Behavioural Science to Support COVID-19 Pandemic Response." *Nature Human Behaviour* 4: 460–71. https://doi.org/10.1038/s41562-020-0884-z.

Varner, Joanna. 2014. "Scientific Outreach: Toward Effective Public Engagement with Biological Science." *BioScience* 64 (4): 333–40. https://doi.org/10.1093/biosci/biu021.
Viciana, Hugo, A. M. Astobiza, A. Fasce, and Ivar R. Hannikainen. 2024. "Scientifically Together, Politically Apart?" *Science & Education*. https://doi.org/10.1007/s11191-024-00587-z.
Weingart, Peter, Marina Joubert, and Kristen Connoway. 2021. "Public Engagement with Science—Origins, Motives and Impact in Academic Literature and Science Policy." *PLOS ONE* 16 (7): e0254201. https://doi.org/10.1371/journal.pone.0254201.
Wellcome Trust. 2024. "Wellcome Global Monitor: How the World Feels about Science and Health." https://wellcome.org/reports/wellcome-global-monitor/2024.
Wilson, Kerrie A., Nancy A. Auerbach, Kadina Sam, et al. 2016. "Conservation Research Is Not Happening Where It Is Most Needed." *PLOS Biology* 14 (3): e1002413. https://doi.org/10.1371/journal.pbio.1002413.
Wintterlin, Florian. 2025. "Improving Trust in Science Through Communication? The Role of Mediating and Moderating Factors." *International Journal of Public Opinion Research*. https://doi.org/10.1093/ijpor/edaf055.

# Chapter 11
# Science Diplomacy

**Cristian Mejia**

## 1 Science Diplomacy and Its Changing Landscape

> How much has changed over 15 years. All the challenges—the environmental, social, and demographic challenges—that were present in 2010 are, 15 years later, much more challenging and much more pressing.—**Mark Walport**

The optimism that once defined science diplomacy has given way to a starker reality. When the AAAS and The Royal Society published *New Frontiers in Science Diplomacy* in 2010, the report articulated a hopeful vision: science as a universal language transcending political divisions (AAAS & The Royal Society, 2010). The framework—*science in diplomacy* (informing policy), *diplomacy for science* (facilitating cooperation), and *science for diplomacy* (building soft power)—reflected an era of faith in globalisation and multilateral institutions.

Fifteen years later, that world has fundamentally transformed. The 2025 report, *Science Diplomacy in an Era of Disruption*, marks a conceptual reckoning (AAAS & The Royal Society, 2025). Great power competition has returned, active conflicts span multiple continents, and the COVID-19 pandemic exposed fragilities in global supply chains. Populist movements have eroded trust in expertise, whilst transformative technologies—from generative AI to CRISPR—evolve faster than governance frameworks can accommodate. Where the 2010 framework emphasised science's neutral convening capacity, contemporary realities demand acknowledging science diplomacy as a *tool* that can advance either collaboration or strategic advantage. Research security and technology competition have become central diplomatic concerns, reflecting what the OECD characterises as growing securitisation of science and technology policy (OECD, 2025).

The 2025 framework by AAAS (AAAS & The Royal Society, 2025) proposes two interactive dimensions replacing the 2010 model's three overlapping

C. Mejia (✉)
Institute for Future Initiatives, The University of Tokyo, Tokyo, Japan



H. Komiyama et al. (eds.), *AI, Innovation, and Beyond*,
https://doi.org/10.1007/978-981-95-7020-1_11

categories. *Science impacting diplomacy* encompasses how scientific evidence informs foreign policy and addresses transnational challenges. *Diplomacy impacting science* acknowledges both enabling and *constraining* roles of diplomatic tools—from facilitating visas to imposing export controls. By explicitly incorporating restriction alongside facilitation, this framework legitimises the strategic calculus governing international scientific exchange: Arctic Council cooperation paused after Russia's invasion of Ukraine, visa protocols tightened over technology transfer concerns, and regional groupings such as the G7, BRICS, and Quad have emerged as primary venues for science and technology coordination (AAAS & The Royal Society, 2025; OECD, 2025). Recent scholarship demands new practices critically assessing when, how, and for whom science diplomacy creates value (Flink, 2021).

The most consequential shift moves from idealism to instrumentality. Science diplomacy can no longer be understood as standing above politics; it is inescapably *of* politics. Yet this recognition need not breed cynicism. Historical precedent demonstrates viability: during the Cold War, scientific exchange persisted through agreements like the 1958 Lacy-Zarubin framework structuring US-Soviet cooperation despite broader hostilities (Reinhardt & Nemudrov, 2023; Nemudrov & Reinhardt, 2023). Smallpox eradication succeeded through concerted international effort when scientific imperatives aligned with national objectives. Contemporary practitioners argue that science diplomacy already functions in competitive contexts; the challenge lies in wielding it effectively (AAAS & The Royal Society, 2025).

> One really important thing about science diplomacy is that it's actually a contact sport.—**Mark Walport**

This pragmatic turn finds expression in Mark Walport's characterisation of science diplomacy as "contact sport"—messy, contingent work of building relationships and negotiating interests when political relations fray. The metaphor acknowledges that effective practice requires direct engagement across differences, strategic acumen alongside ethical commitment, and sustained channels of communication precisely when easier to retreat into isolation. In a disrupted era, such contact has never been more essential.

The sections that follow explore three imperatives: pragmatism in operating within competitive systems whilst seeking collaboration where interests align; equity in moving beyond paternalistic models to build genuine partnerships; and anticipatory governance of emerging technologies before they catalyse conflict (GESDA, 2025; Porsdam & Mann, 2024). Together, these map contested terrain—not nostalgic return to 2010's optimism, but foundation for pragmatic, equitable practice the coming decades demand.

## 2 Security Without Isolation: Building Trust Infrastructure

> Science diplomacy provides an enabling environment and the trust infrastructure on which international cooperation depends.—**Motoko Kotani**

The disrupted landscape generates urgent demand for frameworks sustaining collaboration without compromising security. This tension—between openness essential to scientific progress and protections required in strategic competition—defines a critical frontier. Motoko Kotani's articulation of Japan's approach offers a model navigating this paradox through "security without isolation."

Advanced economies confronting technology transfer risks, intellectual property theft, or dual-use research concerns instinctively tighten controls: visa restrictions multiply, funding becomes nationality-restricted, projects face intensified scrutiny. Whilst addressing legitimate security concerns, these measures risk overcorrection—constraining scientific dynamism they aim to protect. The challenge lies in implementing "strong safeguards with continued international collaboration" rather than allowing research security to become rationale for closing doors (JST, 2024).

"Openness with trust" departs from binary thinking positioning security and collaboration as opposed. Robust integrity mechanisms—research security protocols, data governance standards, transparent evaluation—create conditions enabling meaningful cooperation. Trust becomes infrastructure actively built through institutional design.

Japanese scientific engagement evolved from post-war reconstruction into sophisticated diplomatic asset, leveraging capacity to build partnerships whilst navigating Cold War alignments (Jacoby, 2021). Contemporary frameworks like "Promoting, Protecting, and Partnering Science" balance promoting excellence, protecting intellectual assets, and partnering internationally (JST, 2024), institutionalising "security without isolation" through mutually reinforcing protection and partnership. This trust infrastructure extends to concrete operational capacity: visa facilitation, credential recognition, data-sharing protocols, joint funding instruments, and standardised ethics review—each representing diplomatic negotiation translating into research enablement. Emerging frameworks for quantum technology development similarly demonstrate how science diplomacy can structure international cooperation in sensitive domains whilst managing security concerns (Mafu & Senekane, 2023).

Reconceptualising scientific progress measurement proves diplomatically significant. Traditional metrics—citations, patents, prizes—generate competitive hierarchies. Multi-criteria evaluation (MCE) enables "multi-dimensional data analysis" recognising "diverse national strengths" rather than singular rankings, moving from competition to "co-creation."

MCE frameworks legitimate distinct pathways: nations with modest expenditure but deep traditional and indigenous ecological knowledge contribute vitally to sustainability science without competing on resource-intensive metrics. It enables synergetic collaboration between Global South and Global North both in knowledge production and problem solving. Countries prioritizing domestic challenges—agricultural productivity, water security, disease surveillance—gain recognition for practical contributions beyond publication counts.

This revaluation addresses core resentment: perceptions that existing frameworks validate historically dominant priorities whilst marginalising alternatives. Multiple excellence forms support equitable partnerships where diverse strengths complement rather than rank, aligning with scholarship on decolonising knowledge production and epistemological pluralism (Dajani et al., 2024).

Youth engagement proves central to future viability. Young researchers as "ambassadors" creating "networks without borders and with trust" represent both opportunity and imperative: today's graduate students and early-career researchers will navigate the 2040s and 2050s. Their formative experiences—networks built, norms internalised, collaborations established—shape possibilities for future cooperation.

Contemporary engagement extends beyond exchange programmes. Digital platforms enable collaboration bypassing visa regimes and funding constraints, whilst grassroots movements—climate activism, open science advocacy—demonstrate transnational mobilisation capacity. The EU-Japan Young Leaders Forum exemplifies institutional efforts creating structured multinational dialogues on science-society challenges.

Realising this potential requires institutional reform beyond rhetoric. Existing structures remain gerontocratic, with senior figures dominating whilst younger voices observe. Genuine inclusion demands pathways for early-career researchers shaping agendas, diverse geographic representation, and recognition that younger cohorts possess perspectives on emerging technologies that senior figures may lack. This generational renewal serves equity objectives as Global South youth increasingly possess technical expertise comparable to Global North peers, challenging historical asymmetries.

## 3 Trust vs. Influence: The Science-Policy Paradox

> Our biggest issue is not trust in science; it's the influence of science. We're not out on top in influencing people's behaviours and policies, and that's where we have a lot more competition, a lot more voices, a lot more alternatives.—**Heidi Larson**

Discourse surrounding science and public confidence often presumes a "crisis of trust," citing vaccine hesitancy, climate scepticism, and resistance to expert guidance. Heidi Larson's anthropological research reveals a more troubling reality: trust in science globally remains relatively robust. The genuine crisis lies in declining *influence* of scientific expertise within volatile, hyper-competitive information ecosystems shaping policy and behaviour. This distinction reframes the challenge from restoring confidence to reclaiming relevance.

Research tracking public sentiment across diverse contexts documents that scepticism about specific interventions rarely rejects science's epistemic authority. Rather, publics respond to perceived disconnects between recommendations and lived realities, question whose interests advice serves, or react to how claims are mobilised politically. Publics increasingly find alternative information sources more compelling, relevant, or responsive to their concerns.

Scientific expertise's former gatekeeping advantages—specialised knowledge access, publication control, institutional credibility commanding media deference—have eroded dramatically. Social media enables direct communication bypassing intermediaries. Algorithmic recommendation systems prioritise engagement over accuracy, amplifying emotionally resonant content regardless of epistemic quality. Disinformation campaigns weaponise uncertainty and manufacture controversy where consensus exists (Panter-Brick et al., 2024).

Scientific voices now compete with fundamentally different authority forms: personal testimony authenticity, community identity solidarity, influencer charisma, narrative emotional resonance. Traditional communication strengths—precision, qualification, tentative language acknowledging uncertainty—often function as handicaps. Carefully explained probabilistic findings compete poorly with absolutist oversimplified claims that prove more convincing, shareable, memorable, and persuasive.

Implications prove stark: if science diplomacy's bridging function depends on science's perceived authority, influence erosion undermines foundations. Why prioritise scientific advice when constituencies respond more strongly to other sources? How can Track II diplomacy—unofficial dialogue channels through scientists and academics—function when scientists lack public standing making such exchanges politically consequential? The challenge extends beyond communications strategy to encompass expertise's systemic position within democratic governance.

Rising aggression directed at scientists, documented through *Nature* and *Science* surveys, reveals widespread verbal and physical abuse across disciplines and geographies (Nogrady, 2021, 2022). Consequences extend beyond trauma to systemic self-censorship: scientists withdrawing from public engagement to avoid hostility precisely when scientific voices are most needed for climate adaptation, pandemic preparedness, and technology governance debates.

For science diplomacy, this represents critical vulnerability. Scientists fearing public engagement cannot serve diplomatic bridging functions. Track II diplomacy requires willing participants; soft power requires visible leadership; evidence-to-policy translation demands researchers navigating contentious discourse. Aggression-driven silence atrophies diplomatic functions.

Aggression sources intersect: political polarisation transforms scientific questions into partisan identities; economic disruption generates resentment against perceived elite detachment; digital platforms enable coordinated harassment; governments undermine scientific authority when findings threaten favoured policies.

Cultivating alternative spaces and networks where scientists engage despite hostile domestic contexts proves essential. Strategic transnationalism emerges: whilst digital platforms amplify harassment, they also enable diaspora connections and international collaborations providing psychological support and institutional protection. Science diplomacy must create formal structures supporting transnational engagement, ensuring researchers under pressure have viable alternatives to silence.

Reconceiving scientific communication requires rejecting "dumbing down"—translating complexity into simplified messages presumed accessible. Publics ask not "Can I understand this?" but "Why does this matter to me?" Effective communication requires demonstrating *relevance* rather than *simplicity*.

Making science relevant demands understanding specific audience contexts—anthropological and social scientific research often marginalised in STEM-dominated policy discussions. "Why does this matter" has different answers: Ugandan policymakers centre on economic transformation; European publics on employment security and lifestyle; Pacific Island nations on existential survival under climate change.

Demonstrating relevance requires expanding the "table" for science-policy dialogue. COVID-19 saw scientists appear with political leaders excluding education ministers, business leaders, community representatives, civil society—reinforcing perceptions serving particular interests rather than collective welfare. Inclusive configurations with "changing seats" facilitate engagement of diverse stakeholders and integrate multiple perspectives serving broad publics and bridging divides.

Implications extend beyond communication to institutional design. Audience-specific relevance requires distributed capacity tailoring engagement across contexts rather than centralised messaging. This demands investing in local scientific capacity contextualising global knowledge, supporting translational research addressing community priorities, and creating governance mechanisms incorporating local and indigenous knowledge alongside Western paradigms (Dajani et al., 2024; Panter-Brick et al., 2024).

"There is no rule book right now" for creating appropriate science-policy dialogue "tables." "Nimble and responsive" approaches with "changing seats" tailored to specific issues reflect lessons where standardised protocols failed whilst contextually attuned improvisation succeeded.

This adaptive imperative challenges institutionalisation. Bureaucratic organisations excel at procedures but struggle with improvisation; diplomatic protocols provide predictability but constrain flexibility; accountability demands consistency rather than rewards responsiveness. Yet contemporary volatility, context diversity, and unpredictability of salient scientific issues argue for retaining adaptive capacity.

Adjacent domains offer models: humanitarian response systems with rapid assessment and flexible deployment; adaptive management in environmental governance building learning into implementation; agile development maintaining coherence with uncertainty. Adapting such approaches requires rethinking training programmes, institutional structures, and evaluation metrics valuing responsiveness alongside traditional diplomatic consistency.

## 4 From Aid to Agency: Science as Economic Transformation

> Science for Africa is Africa's new passport into diplomacy, rebranding us from consumers of finished products and suppliers of raw materials, as well as aid recipients, to authentic contributors to scientific knowledge.—**Monica Musenero Masanza**

If the preceding sections have mapped challenges facing science diplomacy in the Global North—security tensions, erosion of influence, institutional adaptation—Monica Musenero Masanza's intervention reframes the entire discourse from the perspective of the Global South. Her critique is fundamental: the dominant narratives of science diplomacy, even when well-intentioned, perpetuate patterns of dependency and marginalisation that render genuine partnership impossible. For African nations and the broader Global South, she argues, science diplomacy cannot be understood as an abstract exercise in bridge-building but must serve as an instrument of economic transformation—a means of escaping the "uncomfortable classifications" of underdevelopment through strategic positioning within emerging technological paradigms.

Musenero Masanza's framing centres economic agency rather than epistemic contribution. Whilst Global North discourse often celebrates international collaboration for advancing knowledge or addressing shared challenges, her emphasis falls squarely on "economic flow of resources" and the imperative that science "translates into wealth and well-being." This prioritisation reflects material realities: for nations struggling with poverty, infrastructure deficits, and limited fiscal capacity, the primary value of science lies not in satisfying intellectual curiosity but in driving socioeconomic development. Science diplomacy, in this context, becomes meaningful only insofar as it facilitates technology transfer, enables market access, builds domestic capacity, and ensures equitable participation in value creation.

Contemporary asymmetries trace to Africa's exclusion during the first three industrial revolutions. Where European and North American economies transformed through mechanisation, electrification, and computerisation, African nations "either completely missed or were improperly positioned to harness" these transitions, resulting in diplomatic participation "unequally yoked"—lacking economic leverage, technological capacity, or institutional infrastructure enabling genuine partnership.

This framing challenges narratives attributing development gaps to governance failures, cultural factors, or insufficient education investment. Instead, it foregrounds systemic exclusion from value creation during critical transitions—maintained through colonial extraction, post-colonial dependency, and contemporary marginalisation. If historical asymmetries resulted from structural position rather than inherent capacity, contemporary initiatives must address power imbalances directly rather than assuming goodwill and knowledge exchange spontaneously generate equity.

Literature increasingly validates this critique. Research documents how collaborative research reproduces extractive patterns through "parachute science"—Northern researchers conducting Southern fieldwork, publishing in Northern journals, accruing rewards whilst Southern collaborators remain marginalised (Dajani et al., 2024). Analysis of Global South science diplomacy highlights increasing agency of nations refusing passive roles, setting foreign policy agendas based on domestic priorities and demanding reciprocal rather than charitable relationships (Ruffini & Krasnyak, 2023).

The fourth and approaching fifth industrial revolutions offer unprecedented repositioning opportunity. "Africa is able to sit and do innovation and science in real time with the rest of the world"—not as technology consumers but as cutting-edge development participants. This temporal simultaneity matters profoundly: when technologies emerge, standards remain fluid, ecosystems form, and early movers shape trajectories rather than merely adapt.

Yet significant asymmetries persist. Advanced computing infrastructure remains concentrated in wealthy nations; venture capital flows overwhelmingly toward Northern startups; patent systems privilege established players; regulatory frameworks reflect Northern rather than Southern realities (OECD, 2025). The opportunity is real but contested—achievable only through deliberate strategy.

Uganda's approach offers a worked example. Kiira Motors—an automotive manufacturer leveraging the "window presented by the green energy revolution"—illustrates using technological transitions to enter industries where incumbency advantages reduce. Electric vehicles require different supply chains and infrastructures than internal combustion automobiles; this disruption creates openings for new entrants securing necessary capabilities. Similarly, investments in "deep tech" and "climate tech" target domains where solutions remain under development, creating potential for innovation rather than mere adoption.

Systematic obscuring of economic models and value chains in emerging technology domains, particularly AI, proves profoundly disempowering for Global South nations. Whilst AI dominates high-level discussions, conversations focus on capabilities, risks, and ethics whilst treating economic dimensions as "amorphous." Northern actors, embedded within ecosystems where economic models are familiar, engage substantively with business implications even when unstated. Southern actors, positioned outside these ecosystems, confront presentations foregrounding technical capabilities whilst leaving business models implicit—unable to assess where value accrues or how to position strategically. This problem extends beyond AI: trade negotiations around intellectual property regimes, data localisation, and technology transfer involve complex economic stakes obscured by technical or security framings.

Addressing this asymmetry requires fundamental transparency about economic stakes. Discussions must "place science side by side with the economic flow of resources," insisting technical cooperation cannot divorce from questions of who benefits, how value distributes, and whether partnerships reinforce or challenge hierarchies. This aligns with scholarship on capacity building and training as prerequisites for meaningful Global South participation, ensuring partnership involves genuine knowledge transfer rather than merely accessing Southern resources whilst reserving value creation for Northern institutions (Höne & Ittelson, 2023).

Science as Africa's "passport into diplomacy" captures transformation from aid recipients to knowledge contributors, from consumers to manufacturers and innovators, from supplicants to partners with valuable capabilities. This requires demonstrating that African nations possess assets wealthier countries need, shifting dynamics from charity to mutual interest.

Such assets include populations increasingly educated in STEM fields offering human capital for collaborative research; access to markets representing over a billion consumers creating commercial incentives; unique contexts—disease burdens, climate conditions, infrastructure constraints—serving as testing grounds for innovations with global applicability; political influence in multilateral forums where numerical weight matters. Each represents potential basis for partnerships structured around complementarity rather than dependency (Gual Soler, 2021).

Yet realising this potential demands building negotiating capacity securing favourable partnership terms, institutional infrastructure absorbing and deploying transferred technologies, regulatory sophistication balancing openness with protecting domestic interests, and political will resisting arrangements offering short-term gains at long-term autonomy cost. The science diplomacy agenda emerging from the Global South increasingly emphasises these capacity-building imperatives, recognising effective partnership requires parties capable of advocating for their interests rather than simply accepting terms set by more powerful actors (Ruffini & Krasnyak, 2023; Ruland et al., 2023).

Musenero Masanza's insistence that partnerships must "uplift our people while shaping a global innovation agenda that is inclusive, sustainable, and fair" establishes multiple criteria for success. Uplift requires tangible improvements in welfare, not merely participation in prestigious collaborations. Shaping the agenda demands voice in determining priorities rather than merely responding to others' initiatives. Inclusivity necessitates genuine access to opportunities rather than token representation. Sustainability implies long-term viability rather than dependence on continued external support. Fairness requires equitable distribution of benefits rather than concentration in already-advantaged nations.

Critically, this reframing rejects "pity" narratives positioning African nations as victims needing rescue, instead emphasising agency, innovation, and leadership. The African Union's Digital Transformation Strategy, Smart Africa initiative, and technology hubs in Lagos, Nairobi, and Cape Town demonstrate sophisticated policy thinking and entrepreneurial ecosystems producing globally competitive innovations. Moving from charity models—aid, remediation, technology transfer on donor terms—to partnership models based on co-investment, mutual learning, and joint priority-setting creates foundations for durable equity rather than entrenching long-term subordination.

For Global South nations, science diplomacy's value lies primarily in contributing to economic transformation. Success requires measuring outcomes not merely in scientific metrics but developmental ones: jobs created, industries established, capabilities enhanced, wealth generated and retained. Science as passport into diplomacy thus articulates aspiration that Global South nations engage from positions of strength derived from scientific capacity, that participation reflects mutual interest rather than charity, and that complementarity-based partnerships generate genuinely shared benefits. Achieving this requires not only building domestic capacity but fundamentally restructuring international scientific engagement terms to ensure equity rather than perpetuating historical asymmetries.

## 5 Science Diplomacy in Reality: How to Make It Effective?

> Science diplomacy is not necessarily a force for good… It is a tool that can be used for good or bad.—**Sudip Parikh**

The preceding sections have explored models and aspirations for science diplomacy in an era of disruption: Japan's trust infrastructure, the challenge of reclaiming scientific influence, and the Global South's demand for economic transformation. Sudip Parikh's realist intervention anchors these aspirations in a starker assessment. Science diplomacy, he argues, cannot be understood as inherently beneficial—a "light" that naturally brings nations together. Rather, it functions as an instrument that can serve collaboration or competition, bridge-building or strategic advantage, depending on how it is wielded and toward what ends. This instrumentalist framing, whilst potentially deflating to idealists, provides the conceptual foundation for effective practice precisely because it demands clear-eyed engagement with power, interests, and structural constraints.

This reflects broader scholarship turning toward realist and critical perspectives. Where early literature portrayed the field through normative lenses—emphasising ought over is, celebrating successes whilst marginalising failures—recent work interrogates whose interests science diplomacy serves, when it functions as cover for strategic competition, and how power asymmetries shape outcomes (Flink, 2021; Ruffini, 2020). Realising potential requires confronting uncomfortable realities about how it operates within competitive international systems.

Parikh describes himself as an "optimist by experience"—someone whose confidence in science diplomacy's possibilities derives not from naive faith but from documented historical achievements. His primary exemplar, smallpox eradication during the Cold War, illustrates that profound international cooperation can occur even amidst intense geopolitical rivalry when scientific imperatives align with broader interests. The campaign succeeded through formal institutional mechanisms, particularly the 1958 Lacy-Zarubin Agreement that structured US-Soviet scientific exchange despite broader antagonisms (Reinhardt & Nemudrov, 2023; Nemudrov & Reinhardt, 2023).

Historical grounding provides pragmatic reference points. "Can you do science diplomacy during a hot war? No, you can't. Can you do science diplomacy at a time with countries that you have very large geopolitical tensions with?… yes, you can." Active military conflict typically suspends even scientific cooperation, whilst rivalry short of warfare permits engagement if properly structured.

Other successes reinforce this: the Montreal Protocol coordinating ozone-depleting substance phase-outs across divergent interests; acid rain mitigation through bilateral cooperation; cancer research extending lifespans through distributed trials. Each demonstrates that when scientific challenges demand collective action and when mechanisms verify commitments and distribute benefits equitably, cooperation persists despite tensions (AAAS & The Royal Society, 2025).

Yet precedents reveal limitations. The Antarctic Treaty System functions where competing sovereignty claims were frozen—pragmatic accommodation rather than resolution (AAAS & The Royal Society, 2010). Arctic cooperation fragmented as climate change transformed the region from scientific commons to strategic resource zone (Ruffin & Ruland, 2022). Arctic Council breakdown following Russia's invasion demonstrates how quickly scientific collaboration becomes collateral damage (AAAS & The Royal Society, 2025). Science diplomacy succeeds when serving mutual interests but proves fragile when those interests diverge or hostilities intensify.

One of the most consequential structural transformations Parikh identifies concerns the rise of private technology corporations as de facto diplomatic actors. Companies such as Alphabet, Microsoft, and Tesla now command research budgets exceeding many national governments, control critical infrastructure including satellite networks and computing capacity, and make strategic decisions with profound geopolitical implications. As Parikh observes, these entities "have their own foreign policies"—they are not merely national champions but supranational actors pursuing distinct corporate interests that may or may not align with any particular nation-state's objectives (AAAS & The Royal Society, 2025).

This development fundamentally challenges the state-centric model that has traditionally defined diplomacy. When a private company controls infrastructure essential to scientific research—cloud computing platforms, AI training capacity, genomic databases—its decisions about access, pricing, and data governance directly shape what research becomes possible and for whom. When technology firms establish research centres globally, they create networks that facilitate knowledge circulation but also enable strategic positioning within national innovation systems. When corporate laboratories produce cutting-edge research, they simultaneously contribute to scientific progress and concentrate commercial advantage (OECD, 2025).

The accountability deficit is profound. Whilst governments remain formally responsible to citizens, multinational corporations answer primarily to shareholders and operate across jurisdictions enabling regulatory arbitrage. They leverage infrastructure chokepoints to extract concessions, shift profits to minimise taxation, and make decisions affecting millions whilst insulated from public oversight. National security implications: when SpaceX's Starlink satellites provide Ukrainian communications infrastructure, a private CEO effectively shapes military operations; when advanced AI development concentrates in few firms, algorithmic governance becomes privatised.

For science diplomacy, this transformation demands reconceptualising who the relevant actors are. Engaging only with national academies and government science advisers whilst ignoring corporate research directors and technology executives misunderstands where scientific capacity increasingly resides. Yet incorporating private actors into diplomatic frameworks raises thorny questions about legitimacy, accountability, and public interest. As Parikh suggests, new governance models are required—mechanisms through which governments can "incentivize" responsible behaviour from firms whilst respecting innovation imperatives and avoiding capture. The changing organisation of biotechnology and bioinnovations, characterised by distributed agencies and increasing Global South participation, demands updated governance approaches acknowledging biosecurity as crucial element of science diplomacy (Namdeo & Zhang, 2024). Absent such models, science diplomacy risks becoming obsolete, addressing formal bilateral relations whilst substantive technological development occurs beyond its purview.

When asked how to manage science diplomacy in domains characterised by both collaborative potential and competitive tension, Parikh invoked the "small yards and high fences" framework employed during recent US administrations. This approach seeks to identify narrowly defined technologies requiring strict export controls and technology transfer restrictions—the "small yards"—whilst maintaining broad openness for scientific exchange in other domains. The metaphor captures a pragmatic middle ground between complete openness, which risks strategic disadvantage through intellectual property theft or dual-use technology transfer, and comprehensive restriction, which would sacrifice the benefits of international collaboration (AAAS & The Royal Society, 2025).

Implementing such frameworks proves extraordinarily complex in practice. Defining which technologies warrant restrictions requires technical expertise, strategic foresight, and political judgment—determining not only current capabilities but anticipating future developments and dual-use applications. Maintaining "small" yards amidst bureaucratic and political pressures toward expansion demands sustained discipline, as security agencies naturally tend toward overcaution whilst affected researchers and institutions advocate for openness. Ensuring that restrictions actually protect strategic advantages rather than merely inconveniencing legitimate collaboration requires sophisticated export control systems and verification mechanisms.

Moreover, alternative frameworks exist, as Parikh acknowledges. Rather than restricting technology transfer, nations might pursue strategies of ubiquitous deployment—ensuring their technology "stacks" become global standards through widespread adoption, thereby gaining influence through infrastructure rather than exclusion. This approach underlies China's Belt and Road technology initiatives and digital infrastructure investments, creating dependencies that provide leverage (OECD, 2025). Neither framework is obviously superior; each involves trade-offs between different forms of strategic advantage, and selection depends on national capabilities, geopolitical position, and political philosophy.

The dialogue at the STS *forum* revealed how these strategic choices play out in practice. When a European audience member noted that the continent had shifted from "open science, open to the world" to "as open as possible and as closed as necessary" following Russia's invasion of Ukraine, he articulated a regional reckoning with the limits of soft power. Science diplomacy's idealist phases may yield to harder-edged calculations when geopolitical realities intrude—a pattern playing out across advanced economies confronting strategic competition with China, technology transfer concerns, and the weaponisation of scientific dependencies. Yet scholarship on science diplomacy between adversarial states demonstrates that maintaining scientific channels even with rivals can yield strategic benefits, enabling limited cooperation on shared interests whilst managing competition elsewhere (Ruland, 2023).

An audience question asked whether and how science diplomacy should be institutionalised as "core capacity for all public science professionals." This query reflects recognition that whilst science diplomacy's importance has grown, dedicated training, career pathways, and institutional structures remain underdeveloped. Most scientists receive no preparation for diplomatic engagement; most diplomats lack sufficient scientific literacy; and systematic evaluation of science diplomacy initiatives remains rare (Flink, 2021; Mauduit & Gual Soler, 2020).

Walport's concluding remarks addressed this directly: "Science diplomacy, if it's to be institutionalised, has got to be institutionalised in a fairly professional way, so that there are good transmission mechanisms between science and policymakers. But that is not something that everyone can do. It is actually a specific activity—science diplomacy." His formulation distinguishes between "scientific engagement"—the broad participation of researchers in international collaboration—and "science diplomacy" proper, which requires specialised skills in navigating political contexts, understanding diplomatic protocols, and translating between scientific and policy communities.

This professionalisation agenda has institutional implications. It suggests need for dedicated training programmes teaching scientists about foreign policy whilst educating diplomats in science and technology. It implies creating career pathways that reward diplomatic engagement rather than treating it as distraction from "real" research. It demands evaluation frameworks assessing science diplomacy's actual impacts rather than merely counting exchanges or agreements. And it requires sustainable funding recognising that effective diplomacy demands long-term relationship building incompatible with short-term project cycles (Höne & Ittelson, 2023).

Yet institutionalisation also risks ossification. Larson's insistence on nimbleness, Musenero Masanza's call for economic relevance, and the audience's emphasis on youth engagement all suggest that overly bureaucratised science diplomacy might sacrifice the adaptability, authenticity, and innovation that make it effective. The challenge becomes building sufficient structure to ensure competence and continuity whilst retaining flexibility to respond to rapidly shifting contexts and emerging opportunities.

## 6 Synthesis: Pragmatism for the Twenty-First Century

The dialogue at the 2025 STS *forum* synthesises into a pragmatic agenda for science diplomacy adequate to contemporary challenges. This agenda rests on several foundational principles derived from the session's collective wisdom:

**Clear-Eyed Realism About Power and Interests** Parikh's insistence that science diplomacy functions as tool rather than inherent good demands acknowledging that it operates within competitive systems where actors pursue national and institutional advantages. This realism need not preclude cooperation but rather provides realistic basis for identifying where interests align sufficiently to enable it.

**Trust Infrastructure as Active Construction** Kotani's framework emphasises that trust does not emerge spontaneously but requires deliberate institutional investment: research security protocols that enable rather than prevent collaboration, standardisation processes that ensure interoperability, exchange programmes that build personal relationships, and verification mechanisms that reduce uncertainty.

**Influence Through Relevance, Not Simplification** Larson's anthropological insight that effective engagement requires demonstrating "why does this matter to me" rather than merely translating complexity into simplicity demands context-specific communication, inclusive dialogue structures, and recognition that different audiences legitimately prioritise distinct values.

**Equity as Precondition for Partnership** Musenero Masanza's economic framing insists that sustainable cooperation requires genuine mutual benefit rather than charity or extraction. For Global South nations, this means ensuring that partnerships build domestic capacity, transfer meaningful technology, and create economic opportunities rather than merely facilitating Northern research agendas.

**Multi-Actor Engagement** Recognising that scientific capacity increasingly resides in private corporations, subnational governments, civil society organisations, and international networks demands expanding science diplomacy beyond traditional state-to-state channels. This diversification enables resilience when official relations deteriorate but also requires new governance mechanisms ensuring accountability.

**Governance as Shared Responsibility** The question of scientific self-governance reveals that neither pure technical autonomy nor complete political control proves adequate. Effective governance of dual-use technologies, AI systems, and biotechnological capabilities requires sustained collaboration between scientists who understand capabilities, policymakers who represent public interests, social scientists who illuminate societal impacts, and affected communities who bear consequences.

**Youth as Agents, Not Subjects** Kotani's emphasis on "young people as ambassadors" reflects recognition that contemporary students and early-career researchers will shape mid-century science diplomacy. Their formative experiences—the networks they build, the norms they internalise, the collaborations they establish—will determine whether the field evolves toward renewed openness or further fragmentation.

**Institutional Capacity with Adaptive Flexibility** Walport's distinction between professional science diplomacy and broad scientific engagement suggests need for dedicated training and career pathways whilst avoiding bureaucratic rigidity. Larson's "no rule book" stance demands retaining nimbleness even as the field professionalises.

## 7 Forward Trajectories: The Coming Decades

Looking beyond the immediate disruptions of the 2020s, several trajectories will likely shape science diplomacy's evolution. The first concerns whether fragmentation into competing technological spheres proves reversible or becomes entrenched. Current trends toward regionalised standard-setting, bifurcated technology ecosystems, and divergent governance frameworks threaten global science's coherence. Yet common challenges—climate adaptation, pandemic preparedness, asteroid defence, ocean governance—generate countervailing pressures toward coordination. Which trajectory prevails will depend substantially on whether science diplomacy can identify and institutionalise sufficient common ground (GESDA, 2025; Porsdam & Mann, 2024).

The second trajectory involves the balance between openness and security. The "growing securitisation" of S&T policy that OECD documents represents response to genuine risks: technology transfer enabling military modernisation, intellectual property theft undermining innovation incentives, dual-use research creating proliferation dangers (OECD, 2025). Yet overcorrection risks sacrificing the diversity of perspectives and transnational collaboration that drive scientific progress. Finding sustainable equilibria—the "security without isolation" that Kotani advocates—will require sophisticated frameworks distinguishing genuine threats from exaggerated fears whilst maintaining scientific dynamism.

The third concerns governance of emerging technologies, particularly artificial intelligence. Existing frameworks—from arms control to environmental treaties—developed over decades through painstaking negotiation and accumulated experience. AI systems evolve on timescales measured in months, deployed globally before governance mechanisms can form. This temporal mismatch between technological change and institutional adaptation generates urgent demand for "anticipatory science diplomacy"—proactive governance of technologies whilst they remain malleable rather than reactive regulation of entrenched systems (GESDA, 2025). Whether this proves achievable remains profoundly uncertain.

The fourth trajectory involves Global South positioning. Musenero Masanza's vision of science as Africa's "passport" articulates aspirations shared across developing nations: moving from periphery to core, from consumption to creation, from supplicant to partner. Realising this vision depends partly on domestic capacity building—Uganda's Kiira Motors illustrates possibilities—but also requires restructuring international frameworks to ensure genuine equity. Whether established powers prove willing to cede space and share benefits, or whether Global South nations must claim position through strategic manoeuvring and coalition-building, will profoundly shape twenty-first-century science diplomacy's character.

The fifth concerns private sector governance. As technological capacity concentrates in multinational corporations operating beyond effective state control, science diplomacy's traditional tools prove inadequate. Developing new mechanisms—whether through international coordination, national regulation with extraterritorial reach, or public-private partnerships—becomes imperative. Yet corporate power and regulatory capture create formidable obstacles. The coming decades will likely witness ongoing contestation over how to ensure that privately developed technologies serve public interests.

## 8 Conclusion

The 2025 STS *forum* dialogue reveals science diplomacy at an inflection point. The disruptions of the 2020s—great power competition, pandemic upheaval, technological acceleration, securitisation of research—have shattered comforting assumptions about science as neutral convening force. Yet precisely because routine diplomatic channels are fraying, scientific networks may provide essential infrastructure for cooperation on challenges that transcend borders: climate destabilisation, pandemic preparedness, AI governance, and emerging technology regulation. The path forward requires abandoning both naive idealism and cynical resignation in favour of pragmatic engagement guided by clear interests and ethical commitments.

Success demands institutional innovation across multiple dimensions: trust infrastructure enabling security without isolation; influence mechanisms demonstrating relevance to diverse publics; equity frameworks ensuring genuine partnership rather than perpetuating dependency; governance models incorporating private sector actors; and professional capacity combining diplomatic skill with scientific literacy. Historical precedents provide guidance but not templates; contemporary challenges differ sufficiently that mechanical application of Cold War models will fail. The future depends substantially on youth—today's students and early-career researchers whose networks, norms, and willingness to engage across divisions will determine whether science diplomacy evolves toward renewed openness or entrenched fragmentation. As Walport's "contact sport" metaphor suggests, effective practice requires not abstract theorising but skilled engagement in spaces where diverse actors meet, negotiate, and occasionally clash—building relationships capable of weathering disagreement whilst identifying sufficient common ground for incremental progress on humanity's shared challenges.

**Competing Interests** No competing interests.

## Session Panelists

**Sir Mark Walport** is Foreign Secretary and Vice President, the Royal Society. He chairs Imperial College Health Partners and Imperial College Academic Health Sciences Centre. He is a non-executive board member of NHS England, and trustee of the British Museum and the Daiwa Anglo-Japanese Foundation. Previous career highlights include: Founding Chief Executive of UK Research and Innovation (UKRI) 2017–2020; Government Chief Scientific Adviser (GCSA) 2013–2017; Member and latterly co-chair of the Prime Minister's Council for Science and Technology; Director of the Wellcome Trust; Professor and Head of the Division of Medicine, Imperial College London; Founder Fellow and first Registrar of the Academy of Medical Sciences; Member of the Advisory Board of Infrastructure UK.

**Dr. Motoko Kotani** is the Executive Director of Science, RIKEN, and the Executive Vice President at Tohoku University, a professor at the Mathematical Institute, and the Principal Investigator at the Advanced Institute for Materials Research (AIMR). She won the 25th Saruhashi Prize for her contributions to Discrete Geometric Analysis on crystal lattices. She has held a number of executive positions, including RIKEN and the past presidency of the Mathematical Society of Japan. Currently, she is a Board of Governors of the Okinawa Institute of Science and Technology and the Science and Technology Co-Advisor to the Minister for Foreign Affairs. In addition, she is the Vice-President for Science and Society of the International Science Council.

**Heidi J. Larson**, Professor of Anthropology, Risk and Decision Science and Director of the Vaccine Confidence Project, London School of Hygiene & Tropical Medicine; Institute for Health Metrics & Evaluation, University of Washington, Seattle. Her research focuses on managing risk and building public trust in the context of pandemics and crises. In 2021, Prof Larson founded the Global Listening Project to investigate public experiences and trust during the COVID-19 pandemic to inform a Societal Preparedness Index for future crises. Prof. Larson previously headed vaccine communication at UNICEF. In 2021 she was awarded the Edinburgh Medal and BBC named her as one of the 100 most influential women in the world.

**Hon. Dr. Monica Musenero Masanza**, DVM, MS, MPH, is a Ugandan scientist & epidemiologist, serving as Minister for Science, Technology, & Innovation, Office of the President. She previously served as Senior Presidential Advisor on Epidemics, leading Uganda's COVID-19 response, and has held key roles in the Ministry of Health, AFENET, and WHO, earning global recognition for her work on Ebola in West Africa. A graduate of Makerere University and Cornell University, she has been pivotal in epidemic preparedness and a pioneer of the One Health movement. As Minister, she champions programmes such as e-Mobility, Pathogen Economy, Industry 4.0+, and Space & Aeronautics, while advancing grassroots innovations and global partnerships to position STI as a driver of Uganda's socio-economic transformation.

**Sudip Parikh** is the CEO of the American Association for the Advancement of Science (AAAS) and executive publisher of the Science family of journals. Parikh has spent over two decades at the intersection of science, policy, and business as a Senate staffer, life science business leader, and in the policy community. He was awarded a National Science Foundation Graduate Research Fellowship while earning his Ph.D. in macromolecular structure and chemistry from the Scripps Research Institute in La Jolla, California and earned his BS in Applied Science from the University of North Carolina at Chapel Hill.

## References

AAAS & The Royal Society. (2010). *New frontiers in science diplomacy*. American Association for the Advancement of Science (AAAS); The Royal Society.

AAAS & The Royal Society. (2025). *Science diplomacy in an era of disruption*. American Association for the Advancement of Science (AAAS); The Royal Society.

Dajani, R., Bayoumi, R., Flowe, H., Rockowitz, S., Stevens, L., & Awad, A. (2024). Neocolonialism and science diplomacy: Lessons learned from the field and the way forward. *Frontiers in Public Health, 12*. https://doi.org/10.3389/fpubh.2024.1472421

Flink, T. (2021). Taking the pulse of science diplomacy and developing practices of valuation. *Science and Public Policy, 49*(2), 191–200. https://doi.org/10.1093/scipol/scab074

GESDA. (2025). *Anticipatory science diplomacy: A framework for global action*. Geneva Science and Diplomacy Anticipator (GESDA). https://www.gesda.global/wp-content/uploads/2025/07/Anticipatory-Science-Diplomacy-A-Framework-for-global-action.pdf

Gual Soler, M. (2021). Science diplomacy in Latin America and the Caribbean: Current landscape, challenges, and future perspectives. *Frontiers in Research Metrics and Analytics, 6*. https://doi.org/10.3389/frma.2021.670001

Höne, K., & Ittelson, P. (2023). Online training and capacity development in science diplomacy: Sharing experiences. *Science Diplomacy Review, 5*(1 & 2). https://www.ris.org.in/sites/default/files/2024-01/SDR_April-and-August-2023.pdf

Jacoby, J. M. (2021). Learning from the earthquake nation: Japanese science diplomacy in the twentieth century. *Journal of Contemporary History, 56*(3), 485–501. https://doi.org/10.1177/00220094211009963

JST. (2024). *The 3rd symposium on science and technology diplomacy for global transformation: Event report*. Japan Science and Technology Agency (JST), Center for Research and Development Strategy (CRDS). https://www.jst.go.jp/crds/sympo/20240321/pdf/20240321_eventreport_en.pdf

Mafu, M., & Senekane, M. (2023). Quantum technology for development framework as a tool for science diplomacy. *Frontiers in Research Metrics and Analytics, 8*. https://doi.org/10.3389/frma.2023.1279376

Mauduit, J., & Gual Soler, M. (2020). Building a science diplomacy curriculum. *Frontiers in Education, 5*. https://doi.org/10.3389/feduc.2020.00138

Namdeo, S. K., & Zhang, J. Y. (2024). Rethinking science diplomacy and global biosecurity: Challenges, emerging practices and the way forward. *International Affairs, 100*(6), 2623–2635. https://doi.org/10.1093/ia/iiae187

Nemudrov, K. A., & Reinhardt, R. O. (2023). Soviet-American scientific and technical cooperation in 1960s–early 1970s: Science diplomacy as a factor of détente. *Nauchnyi Dialog, 12*(3), 414–429. https://doi.org/10.24224/2227-1295-2023-12-3-414-429

Nogrady, B. (2021). 'I hope you die': How the COVID pandemic unleashed attacks on scientists. *Nature, 598*(7880), 250–253. https://doi.org/10.1038/d41586-021-02741-x

Nogrady, B. (2022). 'Overwhelmed by hate': COVID-19 scientists face an avalanche of abuse, survey shows. *Science*. https://doi.org/10.1126/science.abq1755

OECD. (2025). *Science, technology and innovation outlook 2025*. OECD Publishing. https://doi.org/10.1787/5fe57b90-en

Panter-Brick, C., Qtaishat, L., Eggerman, J. J., Thomas, H., Kumar, P., & Dajani, R. (2024). Volunteer programs, empowerment, and life satisfaction in Jordan: Mapping local knowledge and systems change to inform public policy and science diplomacy. *Frontiers in Sociology, 9*. https://doi.org/10.3389/fsoc.2024.1371760

Porsdam, H., & Mann, S. P. (2024). Anticipation and diplomacy (with)in science: Activating the right to science for science diplomacy. *The International Journal of Human Rights, 28*(3), 480–496. https://doi.org/10.1080/13642987.2023.2269102

Reinhardt, R. O., & Nemudrov, K. A. (2023). Lacy-Zarubin agreement of 1958: Origins of Soviet-American science diplomacy. *Nauchnyi Dialog, 12*(1), 431–446. https://doi.org/10.24224/2227-1295-2023-12-1-431-446

Ruffin, N., & Ruland, A.-L. (2022). Between global collaboration and national competition: Unraveling the many faces of Arctic science diplomacy. *Polar Record*, *58*. https://doi.org/10.1017/S0032247422000158

Ruffini, P.-B. (2020). Conceptualizing science diplomacy in the practitioner-driven literature: A critical review. *Humanities & Social Sciences Communications*, *7*(1). https://doi.org/10.1057/s41599-020-00609-5

Ruffini, P.-B., & Krasnyak, O. (2023). Science diplomacy from a nation-state's perspective: A general framing and its application to Global South countries. *Science and Public Policy*, *50*(4), 771–781. https://doi.org/10.1093/scipol/scad023

Ruland, A.-L. (2023). The effectiveness of science diplomacy between adversarial states: Insights from US-Cuban and US-Iranian science collaborations. *The Hague Journal of Diplomacy*, *18*(4), 509–533. https://doi.org/10.1163/1871191x-bja10147

Ruland, A.-L., Rueffin, N., Cramer, K., Ngabonziza, P., Saxena, M., & Skupien, S. (2023). Science diplomacy from the Global South: The case of intergovernmental science organizations. *Science and Public Policy*, *50*(4), 782–793. https://doi.org/10.1093/scipol/scad024

# Concluding Reflections: Science and Technology for the Future of Humankind

**Yousif Elsamani**

## A Moment of Unprecedented Possibility and Profound Fragility

We are living at a historical turning point. Eighty years ago, in 1945, two events helped define the modern scientific age: the creation of the United Nations and the articulation of science as an "endless frontier." Together, they expressed a postwar conviction. Human progress would depend on openness, cooperation, and the steady expansion of knowledge. Today, humanity stands on the threshold of another transformation no less consequential. Artificial intelligence is accelerating discovery across fields. Biological research is moving at an extraordinary speed. Sustainability transitions are reorganizing energy and industrial systems. Scientific capabilities once thought impossible are entering daily life. Yet, this new age of possibility arrives as multilateralism comes under pressure, trust in science is increasingly contested, and social cohesion is fraying.

This paradox is central to the message of this volume: science and technology are already transforming our world, but the critical challenge is whether humanity can direct these forces wisely. Across the book, one consistent lesson emerges: power without direction is dangerous, innovation without governance is unstable, and discovery without social purpose is insufficient. The key takeaway is that it is not enough to generate knowledge; it must also be embedded in effective institutions, guided by values, and contribute to systems that unify rather than fragment humankind.

This is why the present volume matters. On the surface, its chapters span a wide terrain: AI and the future university, decarbonization, nature-positive transitions, brain health, basic science, business transformation, science communication, and science diplomacy. Beneath this diversity lies a deeper coherence. The book is not mainly about technologies themselves. It is about the conditions under which science and technology can remain in service of human dignity, social trust, planetary


H. Komiyama et al. (eds.), *AI, Innovation, and Beyond*,
https://doi.org/10.1007/978-981-95-7020-1

sustainability, and a more just future. In that sense, it is a book about the governance of possibility.

## The Lesson of this Book: Technology Is Not Enough

A first and central lesson of this volume is that technology is never enough by itself. Scientific and technological breakthroughs may open new pathways, but they do not determine their own ends. In chapter after chapter, we see that progress depends on how discoveries are translated through institutions, incentives, infrastructures, values, and forms of collective action.

The chapters on artificial intelligence make this clear from the outset. AI is presented not as a neutral upgrade to existing systems, but as a force that is already reshaping education, research, governance, communication, and culture. In the university context, the issue is not simply whether AI can improve efficiency. Instead, institutions must preserve academic integrity, protect diversity, and ensure equitable access while navigating rapid technological change. Universities are called upon to act not only as adopters of AI, but also as ethical stewards. They serve as public intermediaries, balancing innovation with responsibility.

The sustainability chapters make the same argument in a different register. The path to a zero-carbon society is not a matter of substituting one energy source for another; it is a systems challenge involving infrastructure, markets, finance, geopolitics, regulation, and social coordination. Likewise, the move toward a nature-positive economy cannot be achieved merely by adding biodiversity goals onto existing frameworks. It requires a deeper rethinking of what societies measure, how economies account for value, and how ecosystems are recognized as foundational rather than peripheral. What these chapters show is that the decisive unit of change is rarely the device or the technology itself. It is the system in which technology is embedded.

The same systemic lesson reappears in the chapters on business transformation and innovation. The problem is not a shortage of invention. It is a shortage of transformation. New technologies become game-changing only when they are coupled with viable business models, organizational learning, absorptive capacity, and ecosystems that enable value creation at scale. Invention without institutional redesign remains trapped at the level of possibility. This insight extends the lesson of the sustainability chapters into the economic sphere: what matters is alignment between capability, structure, and purpose.

Even the chapters on communication and diplomacy reinforce this theme. Science communication does not fail because the public lacks information alone; it fails when trust, dialogue, relevance, and coordination are weak. Science diplomacy does not succeed because science stands above politics; it succeeds, when it does, because institutions and actors create sufficient trust infrastructure, reciprocity, and shared incentives to make cooperation durable under real-world conditions. Again, the point is the same: knowledge becomes effective only when its social conditions are cared for.

The book demonstrates that science and technology must be understood as parts of broader human systems. Their impact depends on governance, access, inclusion, and the purposes they serve. The key message: technology is necessary for progress, but never sufficient alone.

## Connecting the Spines of the Porcupine

If technology is not enough, then what is missing? One answer offered throughout this book is connection. In his opening vision for the forum, Chairman Komiyama returned to the image of the porcupinefish, whose spines stand apart from one another as a metaphor for fragmented knowledge. Humanity has accumulated vast stores of scientific, technological, tacit, and cultural understanding, but when these forms of knowledge remain disconnected, they cannot function as a whole. The problem is not simply that we need more expertise. It is that we need better ways to link expertise across domains and direct it toward shared purposes.

This challenge appears across the entire volume. AI forces us to connect computer science with ethics, law, education, labor, and cultural diversity. Sustainability requires integrating climate science, industrial systems, finance, public policy, ecology, and community participation. Brain augmentation demands conversation between neuroscience, medicine, philosophy, law, and human rights. Business transformation requires engineering, organizational design, markets, regulation, and societal purpose to move together. Science communication cannot succeed through isolated expertise alone; it depends on the coordination of researchers, institutions, communities, and communicators. Science diplomacy likewise sits at the intersection of science, foreign policy, development, security, and global justice.

The implication is profound. The great challenges of the twenty-first century are no longer tractable within narrow disciplinary boundaries. They are irreducibly interconnected. Climate change is not merely an environmental issue; it is also an energy, health, finance, governance, and justice issue. AI is not simply a computational issue; it is also an educational, social, geopolitical, linguistic, and ethical issue. The future of humankind will therefore depend not only on excellence within disciplines, but on our ability to connect disciplines, sectors, and institutions in ways that generate coherent action.

The closing plenary sharpened this point by insisting that the future demands not only the knowledge of the natural sciences but also the insights of the social sciences and the wisdom of the humanities. That is not a rhetorical flourish. It is a recognition that scientific capability alone cannot answer questions of purpose, justice, legitimacy, and meaning. Natural science can tell us what is possible. Social science can illuminate incentives, institutions, and behavior. The humanities help us ask what ought to be done, what should be protected, and what kind of civilization we wish to build. Only when these forms of inquiry are linked can science and technology genuinely serve humanity.

In this sense, the "porcupine spines" metaphor is not merely about interdisciplinarity. It is about civilizational intelligence. Humanity now possesses immense knowledge, but the defining challenge is whether that knowledge can be connected across its fragments and directed toward wise ends. The future will belong not to those who merely accumulate more specialized insight, but to those who can integrate knowledge without flattening its diversity and mobilize it without sacrificing human dignity.

## From Knowledge to Wisdom

The closing plenary offered perhaps the most important conceptual key for this volume: data, when organized, becomes information; information, when explained, becomes knowledge. But we now require more than knowledge. We require wisdom.

This distinction lies at the heart of the book. Much of contemporary scientific and technological culture remains oriented toward capability: more computation, faster discovery, greater precision, broader scale, deeper optimization. These achievements matter enormously. Yet the chapters in this volume repeatedly show that knowledge alone does not tell us how power should be used, what limits should be respected, or which futures deserve to be built.

Artificial intelligence illustrates this problem vividly. AI can produce unprecedented forms of inference, automation, and design, but it can also intensify bias, centralize power, increase energy demand, erode critical reasoning, or deepen exclusion if access remains unequal. Neurotechnology can enhance cognition or restore function, but it can also threaten mental privacy and blur the line between care and control. Sustainability tools can help decarbonize systems, but if they are deployed without fairness or political legitimacy, they may provoke backlash and deepen inequality. Communication can spread knowledge, but not necessarily trust. Diplomacy can connect institutions, but not necessarily on equitable terms. In each case, technical capacity opens possibilities that only wisdom can evaluate.

Wisdom, as it emerges from this book, is not an abstract virtue floating above institutions. It has a practical meaning. It is the ability to align power with purpose. It is the ability to hold long-term consequences in view even under conditions of urgency. It is the ability to balance openness with protection, innovation with dignity, and efficiency with justice. It is the discipline of asking not only what can be done, but what should be done, for whom, under what constraints, and at what cost.

The plenary's reflection on the values of science deepens this argument. Science, we were reminded, rests on truth, honor, creativity, openness to the other, tolerance for the contrarian view, civility in discourse, and the disciplined use of evidence, rationality, and logic. These are not merely professional norms for laboratory life. They are values worth defending for a humane society itself. A world that abandons them may still generate information, but it loses the conditions under which knowledge can remain trustworthy, self-correcting, and socially beneficial.

This is why the book's concluding challenge is not simply more innovation, more acceleration, or more technical capacity. It is the cultivation of wisdom across institutions, governance, public discourse, and international cooperation. Wisdom is what allows societies to harness discovery without becoming captive to its momentum. Wisdom is what enables restraint where restraint is needed, courage where courage is needed, and humility where uncertainty is irreducible. Wisdom is what prevents the future from being governed solely by the imperatives of speed, scale, and competition.

## A New Stewardship for the Future of Humankind

If wisdom is the highest requirement, then stewardship is the most fitting word for the task ahead. This volume, taken as a whole, is a call for a new stewardship of science and technology.

It is a call for stewardship of **science**: defending basic research, protecting curiosity-driven inquiry, and preserving the openness that allows science to remain fecund rather than merely instrumental. The closing plenary was right to insist that basic research is the pacemaker of technological progress, and that what matters is not only the finality of answers but the fecundity of questions. A civilization that ceases to ask bold questions will eventually cease to renew itself.

It is a call for stewardship of **institutions**: universities, research systems, communication platforms, diplomatic channels, and governance frameworks capable of translating knowledge into public value. These institutions must now evolve under pressure, without surrendering their deepest purposes. Universities, for example, must remain places of fundamental inquiry even as they become more active stewards of AI, equity, and public trust.

It is a call for stewardship of **nature and the planet**: moving from extractive systems toward circularity, from narrow carbon accounting toward broader ecological responsibility, and from isolated technical fixes toward coordinated transitions that respect planetary limits. Sustainability, as this book repeatedly shows, is not only about reducing harm. It is about redesigning the material basis of civilization in ways that remain livable, just, and resilient.

It is a call for stewardship of **human dignity**: ensuring that AI, neurotechnology, automation, and innovation do not erode autonomy, belonging, or meaning. The most advanced society will not be the one that merely maximizes output, but the one that protects freedom, cultivates critical thought, honors diversity, and expands the conditions for human flourishing.

And it is a call for stewardship of **cooperation**. The conditions for global collaboration are more fragile than they once seemed, but this book offers no excuse for retreat. On the contrary, it suggests that cooperation must now be rebuilt under tension, not postponed until ideal conditions return. Security concerns are real. Competition is real. But so too is the need for trust infrastructure, shared standards, reciprocal partnerships, and multilateral imagination. If science diplomacy, science communication, and global academic collaboration all teach us anything, it is that the alternative to difficult cooperation is not sovereignty in splendid isolation, but fragmentation in a world of shared risks.

The voices of the young leaders in the closing plenary add one final dimension that should not be overlooked. They remind us that the future is not something being discussed only for the next generation; it is already being inherited by them. The warning of an "AI poverty line," the concern for fairness in implementation, the insistence on practical ethics rather than abstract principles, and the call to include diverse voices all underscore a simple truth: intergenerational responsibility is no longer optional. The futures being designed now will define the horizon of possibility for those who live with their consequences longest.

The task ahead, then, is not to slow science and technology, nor to romanticize a past of simpler systems. It is to govern power without suffocating discovery, to preserve openness without naïveté, and to align innovation with humane purpose. This requires courage from governments, responsibility from business, integrity from science, and participation from society. It requires better institutions, better forms of cooperation, and a clearer moral vocabulary for what progress should mean.

The future of humankind will not be determined only by what science and technology make possible. It will be determined by whether we can connect knowledge across its fragments, preserve the values that make inquiry trustworthy, and cultivate the wisdom to use power in the service of life, dignity, and our shared planetary future. That is the work before us. It is also the hope that runs through this book.

GPSR Compliance
The European Union's (EU) General Product Safety Regulation (GPSR) is a set of rules that requires consumer products to be safe and our obligations to ensure this.

If you have any concerns about our products, you can contact us on

ProductSafety@springernature.com

In case Publisher is established outside the EU, the EU authorized representative is:

Springer Nature Customer Service Center GmbH
Europaplatz 3
69115 Heidelberg, Germany

www.ingramcontent.com/pod-product-compliance
Ingram Content Group UK Ltd.
Pitfield, Milton Keynes, MK11 3LW, UK
UKHW020952220926
475831UK00001B/39
* 9 7 8 9 8 1 9 5 7 0 1 9 5 *